Giants of Computing

Gerard O'Regan

Giants of Computing

A Compendium of Select, Pivotal Pioneers

 Springer

Gerard O'Regan
SQC Consulting
Mallow, Cork
Ireland

ISBN 978-1-4471-6226-1   ISBN 978-1-4471-5340-5 (eBook)
DOI 10.1007/978-1-4471-5340-5
Springer London Heidelberg New York Dordrecht

Printed on acid-free paper

Springer is part of Springer Science+Business Media (www.springer.com)

To my siblings-in-law
James, Dean, Araguaci and Alice

Preface

Overview

The objective of this book is to consider a selection of men and women who have made important contributions to the computing field. The goal is to provide brief biographical information on each selected pioneer and to give a concise account of their key contributions to the computing field.

It is clearly not feasible, due to space constraints, to consider all of those who merit inclusion, and the selection chosen inevitably reflects the bias of the author. It is the author's aspiration that the reader will find the selection interesting and will gain an insight into the work and contributions of several pivotal pioneers.

Organization and Features

We discuss a selection of historical figures such as the Hellenistic mathematician Archimedes; the nineteenth-century English mathematician George Boole; the nineteenth-century English mathematician and inventor Charles Babbage; the seventeenth-century German mathematician and inventor Wilhelm Gottfried Leibniz; and Lady Ada Lovelace who worked with Babbage on applications of the Analytic Engine.

The selection of early computer pioneers includes Howard Aiken who developed the Mark I analog computer, John Atanasoff who developed the Atanasoff-Berry Computer, Vannevar Bush who developed the differential analyzer, Tommy Flowers who developed the Colossus computer at Bletchley Park in England, Herman Hollerith who developed tabulating machines in the late nineteenth century, John Mauchly who worked on the ENIAC in the USA, Claude Shannon who showed how Boolean algebra could be applied to the design of digital circuits, John von Neumann who made important contributions to mathematics and early computing, the English mathematician Alan Turing who did important work on theoretical computing and defined the *Turing Test* as a way to judge machine intelligence, Sir Frederick Williams who developed the Williams tube which was used in the development of the Manchester Mark I computer and Konrad Zuse who is considered the *father of the computer* in Germany.

We discuss a selection of those who made important contributions to early commercial computing. These include figures such as Gene Amdahl who was the chief architect for the IBM/360 family of computers and who later set up the Amdahl Corporation, John Backus who developed the FORTRAN programming language at IBM, Gordon Bell who was the architect of several PDP and the VAX 11/780 series of computers at Digital Corporation, Fred Brooks who was the project manager for the IBM/360 project and later wrote an influential book *The Mythical Man Month* describing the challenge of delivering a large software project on time and on budget, Gordon Moore who was one of the co-founders of Intel and proposed Moore's law on the doubling of transistor density and William Shockley who codeveloped the transistor with others at Bell Labs.

There is a selection of those who made important contributions to later commercial computing. This includes figures such as Tim Berners-Lee who invented the World Wide Web, Vint Cerf who was one of the codevelopers of TCP/IP; Edgar Codd who developed relational databases, Don Estridge who led the team that developed the IBM personal computer, Gary Kildall who developed the first operating system for a microprocessor and Richard Stallman who has made important contributions to the *free software movement*.

We discuss a selection of those who have made important contribution to software engineering. These include Dines Bjørner who has made important contributions to VDM and RAISE, Edsger Dijkstra who developed the calculus of weakest preconditions, Tom DeMarco who is one of the developers of structured analysis in the 1970s, Michael Fagan who developed the Fagan Inspection Methodology at IBM in the 1970s, Robert Floyd who did early work on compilers and program verification in the 1960s, C.A.R. Hoare who developed the *quicksort* algorithm and axiomatic semantics, Watts Humphrey who has played a key role in developing maturity models such as the CMM and CMMI and in the management aspects of software projects, Ivar Jacobson who is one of the codevelopers of the Rational Unified Process, David Parnas who has proposed a solid engineering approach to software development and Ed Yourdon who was one of the developers of systems analysis and design methodologies.

We discuss a selection of individuals who have made important contributions to theoretical computing and programming languages. These include Noam Chomsky who has done important work on linguistics and grammars, Alonzo Church who made important contributions to logic and computability and developed the lambda calculus and formulated the *Church-Turing thesis*, James Gosling who is the father of the Java programming language and Grace Murray Hopper, a mathematician and computer pioneer, who worked with Aiken on the Harvard Mark I computer and later developed the COBOL programming language and made important contributions to early commercial computing.

Ken Iverson developed the APL programming language, and his ideas on notation as a tool of thought remain influential; Donald Knuth is considered the father of the analysis of algorithms; Dennis Ritchie codeveloped the C programming language and the UNIX operating system; Dana Scott has made important contributions to logic and to the semantics of programming languages and codeveloped denotational semantics with Christopher Strachey; Bjarne Stroustrup developed the C++ programming language; and Niklaus Wirth developed the Pascal programming language.

We discuss a selection of those who have made important contributions to artificial intelligence. These include the seventeenth-century mathematician René Descartes; John McCarthy is considered the father of AI; Marvin Minsky has made important contributions to artificial intelligence, especially in the areas of learning, knowledge representation, common-sense reasoning, neural networks and computer vision and robot manipulation; John Searle proposed the Chinese Room thought experiment as a rebuttal of strong AI; and Joseph Weizenbaum developed the ELIZA program in the 1960s and became a leading critic of AI.

A selection of computer entrepreneurs is presented. This includes Larry Ellison who is the founder and CEO of the Oracle Corporation, Bill Gates who is the founder and chairman of Microsoft, Steve Jobs who was the founder and CEO of Apple Corporation, Ken Olsen who was the founder and CEO of Digital Corporation and Thomas Watson Sr. and Jr. who were past presidents of IBM.

The selection of pioneers is presented in alphabetical order starting with Aiken and ending with Zuse.

Audience

This book is suitable for computing students who are interested in knowing about the men and women who have shaped the computing field. It will also be of interest to general readers.

Acknowledgements

I am deeply indebted to friends and family who supported my efforts in this endeavour.

Cork, Ireland   Gerard O'Regan
2013

Contents

List of Figures

List of Tables

Chapter 1
Background

1.1 Introduction

Computers are an integral part of modern society and new technology has transformed the world. Communication today may be conducted using text messaging, email, mobile phones and video calls over the Internet using Skype. In the past, communication involved writing letters, sending telegrams or using the home telephones. Today, communication is instantaneous between people, and the new technology has transformed the world into a global village. The developments in computers and information technology have allowed business to be conducted in a global market.

A computer is a programmable electronic device that can process, store and retrieve data. The data is processed by a set of instructions termed a *program*. All computers consist of two basic parts, namely, *hardware* and *software*. The hardware is the physical part of the machine, and a digital computer contains memory for short-term storage of data or instructions, a central processing unit for carrying out arithmetic and logical operations, a control unit responsible for the execution of computer instructions in memory and peripherals that handle the input and output operations. The underlying architecture is referred to as the *von Neumann architecture*. The software is a set of instructions that tells the computer what to do, and it is created by one or more programmers. It differs from hardware in that it is intangible.

The original meaning of the word *computer* referred to someone who carried out calculations rather than an actual machine. The early digital computers built in the 1940s and 1950s were enormous machines consisting of several thousand vacuum tubes.[1] They typically filled a large room or building, but their computational power was a fraction of the power of the computers used today [ORg:12].

---

[1] The Whirlwind Computer (developed in the early 1950s) occupied an entire building. One room contained the operating console consisting of several input and output devices.

G. O'Regan, *Giants of Computing: A Compendium of Select, Pivotal Pioneers*,
DOI 10.1007/978-1-4471-5340-5_1, © Springer-Verlag London 2013

There are two distinct families of computing devices, namely, *digital computers* and the historical *analog computer*. These two types of computer operate on quite different principles, and the earliest computers were analog not digital.

The representation of data in an analog computer reflects the properties of the data that is being modelled. For example, data and numbers may be represented by physical quantities such as electric voltage in an analog computer, whereas in a digital computer a stream of binary digits is used.

A digital computer is a sequential device that generally operates on data one step at a time. The data in a digital computer are represented in binary format, and the binary number system employs just two digits: namely, 0 and 1. A single transistor has two states: i.e. on and off and is used to represent a single binary digit. Several transistors are used to represent larger numbers.

Early computing devices include the slide rule and various mechanical calculators. William Oughtred and others invented the slide rule in 1622, and it allowed multiplication and division to be carried out significantly faster than calculation by hand. Blaise Pascal invented the first mechanical calculator in 1642. It was called the Pascaline, and it could add or subtract two numbers. Multiplication or division could be performed by repeated addition or subtraction.

Leibniz[2] invented a mechanical calculator (called the *Step Reckoner*) in 1672. It was the first calculator that could perform all four arithmetic operations: i.e. addition, subtraction, multiplication and division.

James Thompson (who was the brother of Lord Kelvin) did early work on analog computation in the nineteenth century. He invented a wheel-and-disc integrator, which was used in mechanical analog devices, and he worked with Kelvin to construct a device to perform the integration of a product of two functions.

The operations in an analog computer are performed in parallel and they are useful in simulating dynamic systems. They have been applied to flight simulation, nuclear power plants and industrial chemical processes.

Vannevar Bush and others developed the first large-scale general-purpose mechanical analog computer at the Massachusetts Institute of Technology in the late 1920s. Bush's differential analyzer was a designed to solve 6th-order differential equations by integration using wheel-and-disc mechanisms to perform the integration. It allowed integration and differential equation problems to be solved more rapidly.

It consisted of wheels, discs, shafts and gears to perform the calculations and required a considerable effort to be set up by technicians to solve a particular equation. It contained 150 motors and miles of wires connecting relays and vacuum tubes.

Data representation in an analog computer is compact but may be subject to corruption with noise. A single capacitor can store one continuous variable in an analog computer, whereas several transistors are required to represent a variable in a digital computer. Analog computers were replaced by digital computers after the Second World War (Fig. 1.1).

---

[2]Leibniz is credited (along with Newton) with the development of the calculus.

Fig. 1.1 Differential analyzer at Moore School of Engineering, University of Pennsylvania

1.2 Digital Computers

Early digital computers used vacuum tubes to store binary information. A vacuum tube could represent the binary value "0" or "1". However, the tubes were large and bulky and generated a significant amount of heat. This led to problems with their reliability, and air-conditioning was employed to cool the machine.

Shockley and others invented the transistor in the 1950s, and they replaced vacuum tubes. Transistors are small and require very little power. The resulting machines were smaller, faster and more reliable.

Integrated circuits were introduced in the 1960s and a massive amount of computational power may be placed in a very small chip. They are small with little power consumed and may be mass-produced to very high quality standard. Billions of transistors may be placed on an integrated circuit.

The development of the microprocessor allowed a single chip to contain all of the components of a computer from the CPU and memory to input and output controls. The microprocessor could fit into the palm of the hand, whereas the early computers filled an entire room.

The fundamental architecture of a computer has remained basically the same since von Neumann and others proposed it in the 1940s. It includes a central processing unit, the control unit, the arithmetic logic unit, an input and output unit and memory.

1.3 Hardware and Software

Hardware is the physical part of the machine. It is tangible and may be seen and touched. It includes the historical punched cards and vacuum tubes, transistors and circuit boards, integrated circuits and microprocessors. The hardware of a personal computer includes a keyboard, network cards, a mouse, a DVD drive, hard disc drive, printers, scanners and so on.

Software is intangible in that it is not physical, and instead it consists of a set of instructions that tells the computer what to do. It is an intellectual creation of a programmer or a team of programmers. There are several types of software such as system software and application software.

The *system software* manages the hardware and resources and acts as an intermediary between the application programs and the computer hardware. This includes the UNIX operating system, the various Microsoft Windows operating systems and the Mac operating system. There are also operating systems for mobile phones, video games and so on. *Application software* is designed to perform a specific application such as banking, insurance or accounting.

Early software consisted of instructions in machine code that could be immediately executed by the computer. These programs were difficult to read and debug. This led to assembly languages that represented a particular machine code instruction by a mnemonic and the assembly language was translated into machine code by an assembler. Assembly languages were an improvement on machine code but were still difficult to use. This led to the development of high-level programming languages (e.g. FORTRAN and COBOL) where a program was written in the high-level language and compiled to the particular code of the target machine.

1.4 Giants of Computing

The objective of this book is to give a concise account of the work of a selection of those who have made important contributions to the computing field. It is not feasible, due to space constraints, to consider all those who merit inclusion, and the selection chosen inevitably reflects the bias of the author.

The selection is presented in alphabetical order, and approximately 60 individuals are discussed. The account of each pivotal pioneer includes brief biographical information and a concise account of their contribution.

We discuss a selection of historical individuals who provided the foundation for the field. These include George Boole who was an English mathematician who developed the Boolean algebra, which was later recognized by Claude Shannon as providing the appropriate mathematical model for the design of digital circuits. Charles Babbage was an English mathematician and inventor, who designed the Difference Engine and the Analytic Engine. The difference engine was designed to compute polynomial functions and was used for the production of mathematical tables. The design of the analytic engine was essentially the design of the world's first mechanical computer. Lady Ada Lovelace was the daughter of Lord Byron, and she worked with Babbage on applications of the Analytic Engine. She is considered the world's first programmer. Gottfried Wilhelm Leibniz was a seventeenth-century German mathematician and inventor. He developed the binary number system which is fundamental in computing, and he also developed a mechanical calculator for performing elementary arithmetic operations such as addition and division.

Hermann Hollerith developed a tabulating machine to process the 1890 census in the United States, and he later founded a tabulating company that would become International Business Machines (IBM). Vannevar Bush developed the first large-scale general-purpose mechanical analog computer at the Massachusetts Institute of Technology in the late 1920s, and the machine was used to solve differential equations. Claude Shannon was one of Bush's students at MIT, and he showed how Boolean algebra could be applied to the design of digital circuits. The early computer pioneers built analog and digital computers.

Howard Aiken designed and built the Harvard Mark I computer which was an electromechanical calculator. John Atanasoff designed and built the Atanasoff-Berry Computer (ABC) digital computer in 1942. Tommy Flowers designed the Colossus computer at Bletchley Park in England in 1944, and this machine was invaluable in cracking the Lorenz codes during the Second World War. John Mauchly designed the ENIAC and EDVAC. John von Neumann made fundamental contributions to mathematics and gave his name to the fundamental architecture underlying computer systems. Alan Turing did important work in theoretical computing and on early computers at Bletchley Park, and he later designed the NPL ACE computer. Sir Frederick Williams designed the Williams Tube, which was used on the Manchester Mark I computer. Konrad Zuse is considered to be the "father of the computer" in Germany.

We discuss a selection of those who made important contributions to early commercial computing, including Gene Amdahl, who did important work on early IBM computers. He was the chief designer of the IBM S/360 series of computers, and he later formed Amdahl Corporation which became a rival to IBM in the mainframe market. John Backus made important contributions to programming languages, and he was the designer of the FORTRAN programming language. Gordon Bell was the architect for the PDP-4 and PDP-6 and VAX series of computers at Digital Corporation. Fred Brooks was the project manager for the IBM 360 project, and he later wrote the influential book *The Mythical Man Month*. This book describes the challenge of delivering a large project in which software is a major constituent on time, on budget and with the right quality. Gordon Moore was one of the co-founders of the Intel Corporation, and he also formulated Moore's law. William Shockley and others developed the transistor at Bell Labs in the early 1950s.

We discuss a selection of individuals who made important contributions to later commercial computing. Vint Cerf and Bob Kahn invented the transmission control protocol (TCP) in the early 1970s. Edgar Codd developed relational databases at IBM, and these are widely used today. Don Estridge was the project manager for the team that developed the IBM personal computer. Gary Kildall wrote the first programs for the Intel 4004 microprocessor, and he developed the CP/M operating system for microprocessors. Tim Berners-Lee invented the World Wide Web, which has led to a revolution in the computing field.

Software engineering is concerned with the sound engineering of software and is concerned with the design and development of software systems. It includes figures such as Robert Floyd who did important work on parsing and compilers in the 1960s, and he was one of the pioneers in investigating methods to prove the correctness of programs. C.A.R. Hoare has made fundamental contributions to computing, including the development of the quicksort algorithm, the development of axiomatic semantics of programming languages and the Calculus of Sequential Processes (CSP). Dines Bjørner and Cliff Jones developed the Vienna Development Method (VDM) at the IBM laboratory in Vienna. VDM is a method for the formal specification and development of software. Edsger Dijkstra contributed to the development of graph algorithms and to the development of the Algol 60 programming language. His calculus of weakest preconditions is a methodology to develop a program and its proof of correctness hand in hand.

Tom DeMarco has made important contributions to project management and was one of the developers of structured analysis in the 1980s. Michael Fagan developed the Fagan Inspection methodology at IBM, and this methodology is useful in building quality into software. Watts Humphrey made important contributions to software quality and to the development of process maturity models such as the Capability Maturity Model (CMM), PSP and TSP. Ivar Jacobson has made important contributions to UML and to the Rational Unified Process. David Parnas has made important contributions to the software field, and his ideas on the specification, design, implementation, maintenance and documentation of computer software remain important. He advocates a solid classical engineering approach to the development of software. Ed Yourdon has made contributions to systems analysis and design methodologies.

A selection of individuals who have made important contributions to theoretical computing and programming languages is considered. These include Noam Chomsky who did important work on linguistics and grammars. Alonzo Church made important contributions to logic and computability, and he developed the lambda calculus. James Gosling developed the Java programming language at Sun Microsystems. Grace Murray Hopper was a mathematician and computer pioneer who worked with Aiken on the Harvard Mark I. She developed the COBOL programming language. Kenneth Iverson developed the APL programming language. Donald Knuth is considered to be the father of the analysis of algorithms, and he has published the ultimate guide to program development in his massive four-volume work *The Art of Computer Programming*. He also developed the TeX and METAFONT typesetting systems. Dennis Ritchie developed the C programming language and codeveloped the UNIX operating system with Ken Thompson. Dana Scott made important contributions to the semantics of programming language, and codeveloped the Scott-Strachey approach with Christopher Strachey. Bjarne Stroustrup designed the C++ programming language and Niklaus Wirth designed the Pascal programming language. Richard Stallman is the founder of the free software movement with the GNU project.

The field of Artificial Intelligence showed potential in its early days. It includes individuals such as the mathematician and philosopher René Descartes, who formulated Cartesian dualism and made a clear distinction between mind and body. John McCarthy is considered the father of AI, and he believed that common sense knowledge and reasoning can be formalized with logic, and logical reasoning solves common sense problems. Common sense includes basic facts about events, beliefs, actions, knowledge and desires. It also includes basic facts about objects and their properties.

Marvin Minsky is one of the founders of the artificial intelligence field and was one of the co-founders (along with John McCarthy) of MIT's AI laboratory in the early 1960s. He advocates a symbol manipulation approach as the centre of any attempt to understand intelligence. John Searle formulated the Chinese Room thought experiment which is a rebuttal of strong AI. Joseph Weizenbaum developed the ELIZA program, at MIT in 1966, and this famous program interacted with a user sitting at an electric typewriter in the manner of a psychotherapist. It convinced several users that it had real understanding, and this led to users unburdening themselves in long computer sessions. Its success in convincing naïve users deeply disturbed Weizenbaum, and he became an advocate of social responsibility in science and a leading critic of AI research.

A selection of computer entrepreneurs is presented. This includes Larry Ellison who founded Oracle Corporation. Bill Gates founded Microsoft Corporation, Steve Jobs founded Apple Corporation, Ken Olsen founded Digital Corporation and Thomas Watson Sr. and Jr. were the former presidents of IBM.

Chapter 2
Howard Aiken

Howard Aiken made several important contributions to the early computing field. He showed that a large calculating machine could be built that would provide speedy solutions to mathematical problems. He also made important contributions to early computer science education (Fig. 2.1).

He was born in New Jersey in 1900 and grew up in Indiana. He studied electrical engineering at the University of Wisconsin and obtained a bachelor's degree in 1923. He then worked as chief engineer for several years at the Madison Gas Company before taking a position as general engineer in 1927 at the Westinghouse Electric and Manufacturing Company. He was involved in product design and in the design of power plants, and in 1931 he became a district manager at the Line Material Company.

He resigned in 1933 to return to academia to pursue graduate studies in physics. He pursued a Ph.D. degree at Harvard University, and his Ph.D. in Physics was awarded in 1939. He became associate professor of applied mathematics in 1941 and became a full professor in 1946.

It was during his graduate studies that Aiken became conscious of the need for a machine that could deal with many of the tedious calculations that arose in solving differential equations by numerical means. This led him to investigate machines to ease the calculation burden of his research.

2.1 Harvard Mark I

He did some research on what a scientific calculating machine should do and published a report. His idea was to construct an electromechanical machine that could perform mathematical operations quickly and efficiently, and the machine would need to be able to handle positive and negative numbers, scientific functions such as logarithms and be able to work with minimal human intervention.

Fig. 2.1 Howard Aiken

He discussed the idea with colleagues and IBM and was successful in obtaining IBM funding to build the machine. The machine was built at the IBM laboratories at Endicott with several IBM engineers involved in its construction.

The machine became known as the Harvard Mark I (also known as the *IBM Automatic Sequence Controlled Calculator* (ASCC)). Aiken was influenced by Babbage's ideas on the design of the Analytic Engine. The construction took 7 years and was completed in 1943. IBM presented it to Harvard University in 1944, and the machine was essentially an electromechanical calculator that could perform large computations automatically. It could perform addition, subtraction, multiplication, division and refer to previous results.

The Mark I was designed to assist in the numerical computation of differential equations and was 50 ft long, 6 ft high and weighed 5 tonnes. It performed additions in less than a second, multiplications in 6 s and division in about 12 s. It used electromechanical relays to perform the calculations (Fig. 2.2).

It could execute long computations automatically. It used 500 miles of wiring and had over 700,000 components. It was the industry's largest electromechanical calculator, and it had 60 sets of 24 switches for manual data entry. It could store 72 numbers, each 23 decimal digits long. Punched cards were used to input the data, and the results were on either punched cards or an electric typewriter.

It was used by the US Navy for ballistic calculations and remained in use until 1959. The machine cost approximately half a million dollars but was never mass-produced by IBM.

The announcement of the Harvard Mark I led to tension between Aiken and IBM, as Aiken announced himself as the sole inventor without acknowledging the important role played by IBM.

Fig. 2.2 Harvard Mark I (IBM ASCC) (Photo Public Domain)

2.2 Later Work

Aiken also designed and developed the Harvard Mark II, III and IV. The Mark II was funded by the US Navy and completed in 1947. It was faster than the Mark I and employed an electrical memory.

His Mark III (or Aiken Dahlgren Electronic Calculator (ADEC)) was built in 1949 for the Navy and was one of the earliest machines to use magnetic drums for data storage. His Mark IV was completed in 1952 with funding from the US air force.

Aiken also made important contributions to computer science education, and he started one of the earliest computer science academic programs in the world at Harvard.

He received the IEEE Edison Medal in 1970 for his pioneering contributions to the development and applications of large-scale digital computers and to important contributions to education in the computer field.

He retired from Harvard in 1961 and became a distinguished professor of Information at the University of Miami. He became an entrepreneur and founded a New York-based consulting company that focused on assisting ailing companies back to good financial health. He was a consultant to Lockheed Missiles and Monsanto. He died in Missouri in 1973.

Chapter 3
Gene Amdahl

Gene Amdahl is an American computer scientist and entrepreneur. He was the founder of Amdahl Corporation, which became a major rival to IBM in the mainframe market during the 1970s and 1980s.

He was born in South Dakota in 1922, and he served in the American Navy during the Second World War. He obtained a degree in engineering physics from South Dakota University in 1948, and earned a Ph.D. in theoretical physics in 1952 from the University of Wisconsin-Madison. His Ph.D. thesis detailed the design of his first computer, the *Wisconsin Integrally Sequential Computer* (WISC), and the machine was built during the period 1951–1954 (Fig. 3.1).

He joined IBM in 1952 and became the chief designer of the IBM 704 computer, which was released in 1954. Amdahl and Backus[1] designed this large computer, and it was used for scientific and commercial applications. Over 140 of these machines were sold, and the machine was a commercial success for IBM.

Amdahl became the chief architect for the IBM System/360 family of mainframe computers. This family was introduced in 1964, and the IBM chairman, Thomas Watson, called the event the most important product announcement in the company's history. Fred Brooks[2] was the IBM System/360 project manager.

The IBM System/360 was a family of small to large computers, and it offered a choice of 5 processors and 19 combinations of power, speed and memory. There were 14 models in the family. The concept of a *family of computers* was a *paradigm shift* away from the traditional *one size fits all philosophy of the computer industry*, as up until then, every computer model was designed independently.

The family of computers ranged from minicomputers with 24 kb of memory to supercomputers for US missile defence systems. However, all these computers

---

[1] John Backus is discussed in a later chapter.

[2] Fred Brooks wrote an influential paper *The Mythical Man Month* based on his experience as project manager for the IBM System/360 project. He is discussed in a later chapter.

G. O'Regan, *Giants of Computing: A Compendium of Select, Pivotal Pioneers*,
DOI 10.1007/978-1-4471-5340-5_3, © Springer-Verlag London 2013

Fig. 3.1 Gene Amdahl
(Photo Courtesy of Perry
Kivolowitz)

Fig. 3.2 IBM System/360 Model 30 (Courtesy of IBM Archives)

employed the same user instruction set, and the main difference was that for the larger computers the more complex machine instructions were implemented with hardware, whereas the smaller machines used micro code. A customer could start with a small member of the IBM System/360 family and upgrade over time in to a larger computer in the family. This helped to make computers more affordable for businesses and stimulated growth in computer use.

The IBM System/360 was used extensively in the Apollo mission to place man on the moon. The contribution by IBM computers and personnel were essential to the success of the project. IBM invested over $5 billion in the design and development of the IBM System/360. However, the gamble paid off and it was a very successful product line for the company (Fig. 3.2).

Amdahl was appointed an IBM fellow in 1965 in recognition of his contribution to IBM, and he was appointed director of IBM's Advanced Computing Systems (ACS) Laboratory in California and given freedom to pursue his own research projects.

3.1 Amdahl Corporation

He resigned from IBM in 1970 following disagreements with IBM on his ideas for future computer development. This led him to set up his own computer company, Amdahl Corporation, in California. This company was later to become a major competitor to IBM in the mainframe market, although initially Amdahl had difficulty in raising sufficient capital to develop its business. Amdahl received funding from Fujitsu who were interested in a joint development program and also from Nixdorf who were interested in representing Amdahl in Europe. Amdahl computers were compatible with IBM computers and software, but delivered an improved performance at a lower cost. Customers could run IBM System/360 and S/370 applications on Amdahl hardware without buying IBM hardware.

Amdahl launched its first product, the Amdahl 470, in 1975. Its first customer was NASA's Goddard Spaceflight Centre in New York and further sales to universities and other institutions followed. The Amdahl 470 competed directly against the IBM System 370 family of mainframes, which was released in 1972. It was compatible with IBM hardware and software but cheaper than the IBM product: i.e. the Amdahl machines provided better performance for less money.

IBM's machines were water-cooled, while Amdahl's were air-cooled, which decreased installation costs significantly. The use of large-scale integration (LSI) with many integrated circuits on each chip meant the Amdahl 470 was one-third of the size of IBM's 360. Further, its performance was over twice as fast and sold for about 10 % less than the IBM System/360.

Machine sales were slow initially due to concerns over the company's long-term survival. It had sold 50 units by 1977 and it began to pose a serious challenge to IBM in large-scale computer placements. Amdahl decided to sell rather than lease its equipment, and this led to a price war with IBM.

Customers voted with their feet and chose Amdahl as their supplier, and it was clear that Amdahl was a major threat to IBM in the high-end mainframe market. It had a 24 % market share and annual revenues of $2 billion by 1989.

Amdahl worked closely with Fujitsu to improve circuit design, and Fujitsu's influence on the company increased following Gene Amdahl's departure from the company in 1979. However, by the late 1990s it was clear that Amdahl could not compete against IBM's 64-bit zSeries, as Amdahl estimated that it would take $1 billion to create an IBM-compatible 64-bit system. Further, there were declining sales in the mainframe market due to the popularity of personal computers. Amdahl became a wholly owned subsidiary of Fujitsu in 1997, and its headquarters are in California.

3.2   Amdahl's Law

Amdahl formulated *Amdahl's law*, which states a fundamental limitation of parallel computing [Amd:67]. This law expresses the maximum expected improvement to an overall system when only part of the system is improved. The law is used in parallel computing to predict the theoretical maximum increase in speed using multiple processors.

The law is concerned with the expected improvement in a computation that affects a proportion P of that computation and where the improvement has an improvement speed-up of S.

$$\frac{1}{(1 - P) + \frac{P}{S}}$$

The law expresses the theoretical improvement speed-up that may be achieved from introducing parallelism into a program. If P is the proportion of the program that may be made parallel, and $1 - P$ the proportion that must remain serial, then the maximum improvement speed-up that may be achieved by using N processors is given by

$$S(N) = \frac{1}{(1 - P) + \frac{P}{S}}$$

The theoretical improvement from the introduction of a large number of processors tends to the limit $1/(1 - P)$, and an important part of parallel programming is to investigate ways to reduce $(1 - P)$ to the smallest possible value.

Amdahl resigned from Amdahl Corporation in 1979 to pursue other interests. He set up a new company called Trilogy Systems with the goal of creating an integrated chip for cheaper mainframes. The company later worked on very-large-scale integration (VLSI) with the goal of creating a supercomputer to rival IBM. However, Trilogy was not successful and the company later merged with Elxsi in 1985. The merged company also foundered and Amdahl left in 1989, and he later set up two new companies namely Andor Systems in 1987 and Commercial Data Servers (CDS) in 1994.

Amdahl has received various awards in recognition of his contributions to computing. He received the ACM/IEEE Eckert-Mauchly award in 1987 and the IEEE Computer Entrepreneur Award in 1989. He is a fellow of the National Academy of Engineering and IEEE and a distinguished fellow of the British Computer Society.

Chapter 4
Archimedes

Archimedes was a Hellenistic mathematician, astronomer, inventor and engineer. He was born in Syracuse[1] around 287 B.C., and he was a leading scientist and inventor in the Greco-Roman world. He is credited with designing several innovative machines.

His inventions include the *Archimedes Screw* which was a screw pump that is still used today in pumping liquids and solids. The *Archimedes Claw* is claimed to be another of his inventions, and this defensive weapon was designed to protect the city of Syracuse from an attack by an enemy ship. It was also termed the *ship shaker*, and it is believed to have consisted of a crane arm from which a large metal hook was suspended. The claw would swing up and drop down on the attacking ship. It would then lift the ship out of the water and possibly sink it.

Another of his inventions is claimed to be the *Archimedes Heat Ray*. This device is said to have consisted of a number of mirrors that allowed sunlight to be focused on an enemy ship thereby causing it to go on fire. This experiment was replicated by a group of students from Massachusetts Institute of Technology in 2005 with inconclusive results (Fig. 4.1).

There is a well-known anecdote concerning Archimedes and the crown of King Hiero II. The king wished to determine whether his new crown was made entirely of solid gold and to be absolutely certain that the goldsmith had not added a substitute metal such as silver. Archimedes was required to solve the problem without damaging the crown, and as he was taking a bath, he realized that if the crown was placed in water, the water displaced would give him the volume of the crown. From this he could then determine the density of the crown and therefore whether it consisted entirely of gold.

---

[1]Syracuse is located on the island of Sicily in Southern Italy. It was an independent Greek city state in the third century B.C.

G. O'Regan, *Giants of Computing: A Compendium of Select, Pivotal Pioneers*,
DOI 10.1007/978-1-4471-5340-5_4, © Springer-Verlag London 2013

Fig. 4.1 Archimedes in
thought by Fetti

Archimedes is likely to have used a more accurate method than that described
in the anecdote. He had discovered the law of buoyancy known as Archimedes's
principle:

The buoyancy force is equal to the weight of the displaced fluid.

He is believed to have discovered this principle while sitting in his bath and was
so overwhelmed with his discovery that he rushed out onto the streets of Syracuse
shouting *Eureka*.[2] Unfortunately, he forgot to put on his clothes to announce the
discovery. The term *Eureka moment* is used today to refer to a moment when a
scientist has a sudden flash of inspiration, leading to a new invention or the solution
of a particular problem.

The weight of the displaced liquid will be proportional to the volume of the
displaced liquid. Therefore, if two objects have the same mass, the one with
greater volume (or smaller density) has greater buoyancy. An object will float if
its buoyancy force (i.e. the weight of liquid displaced) exceeds the downward force
of gravity (i.e. its weight). If the object has exactly the same density as the liquid,
then it will stay still, neither sinking nor floating upwards.

For example, a rock is generally a very dense material and will generally not
displace its own weight. Therefore, a rock will sink to the bottom as the downward
weight exceeds the buoyancy weight. However, if the weight of the object is less
than the liquid it would displace, then it floats at a level where it displaces the same
weight of liquid as the weight of the object.

His method of determining whether the crown was made entirely of gold may
have employed a scale, where the crown would have been balanced by the equivalent
weight of gold, and then immersed these in water. Any difference in density between
the crown and the gold would have caused the scales to tilt due to differences in
buoyancy.

---

[2]The Greek word "Eureka" (ε ρηκα) means "I have found it".

Archimedes also made good contributions to mathematics including a good approximation to π; contributions to the positional numbering system, geometric series and the calculation of the area under the arc of a parabola; and contributions to mathematical physics.

He also solved several interesting problems: e.g. the calculation of the composition of cattle in the herd of the Sun god by solving a number of simultaneous Diophantine equations. The herd consisted of bulls and cows with one part of the herd consisting of white, second part black, third spotted and the fourth brown. Various constraints were then expressed in Diophantine equations and the problem was to determine the precise composition of the herd. Diophantine equations are named after Diophantus of Alexandria who worked on number theory in the third century A.D.

Archimedes also calculated an upper bound of the number of grains of sands in the known universe. The largest number in common use at the time was a myriad myriad (100 million), where a myriad is 10,000. Archimedes' numbering system goes up to $8*10^{16}$, and he also developed the laws of exponents: i.e. $10^a 10^b = 10^{a+b}$. His calculation of the upper bound includes not only the grains of sand on each beach but on the earth filled with sand and the known universe filled with sand. His final estimate of the upper bound for the number of grains of sand in a filled universe was 10^{64}.

It is possible that he may have developed the odometer,[3] and this instrument could calculate the total distance travelled on a journey. An odometer is described by the Roman engineer Vitruvius around 25 B.C. It employed a wheel with a diameter of 4 ft (i.e. the circumference of the wheel is $\pi*4 = 12.56$ ft), and the wheel turned 400 times in every mile.[4] The device included gears and pebbles and a 400-tooth cogwheel that turned once every mile and caused one pebble to drop into a box. The total distance travelled was determined by counting the pebbles in the box.

He died during the Roman siege of Syracuse around 212 B.C., and a sculpture of a sphere and a cylinder were placed on his tomb. Archimedes had shown that a sphere that is circumscribed by a cylinder with the same diameter and height has 2/3 of the volume and 2/3 of the area of the circumscribed cylinder.

There is a crater on the moon named after Archimedes and the asteroid 3,600 Archimedes is also named after him. The word *Eureka* has become a popular way to express a creative or innovative moment where a flash of inspiration leads to the solution to a particular problem.

---

[3]The origin of the word "odometer" is from the Greek words "οδος" (meaning journey) and "μετρον" meaning (measure).

[4]The figures given here are for the distance of one Roman mile which is shorter than a standard mile on the Imperial system. An Imperial mile is 5,280 ft, whereas the Roman mile is 5,024 ft (i.e. $400*\pi*4$).

Chapter 5
John Atanasoff

John Vincent Atanasoff was born in New York in 1903. He studied electrical engineering at the University of Florida and did a Master's in mathematics at Iowa State College. He earned a Ph.D. in theoretical physics from the University of Wisconsin in 1930 and became an assistant professor at Iowa State College, where he taught mathematics and physics (Fig. 5.1).

He became interested in developing faster methods of computation while doing his Ph.D. research, as he wished to ease the time-consuming burden of calculation. He did some work on an analog calculator in 1936 but concluded that analog devices were too restrictive and could not give him the desired accuracy. His goal was to mechanize calculation to enable computation to be carried out faster.

The existing computing devices were mechanical, electromechanical or analog. Atanasoff developed the concept of a digital machine to perform faster computation in the late 1930s. He believed that his proposed machine offered advantages over the slower and less accurate analog machines. He published the design of a machine to solve linear equations using his own version of Gaussian elimination in the summer of 1939. He then used his research grant of $650 to build the Atanasoff-Berry Computer (ABC), with the assistance of his graduate student, Clifford Berry, from 1939 to 1942.

The ABC was approximately the size of a large desk and had approximately 270 vacuum tubes. Two hundred and ten tubes controlled the arithmetic unit, 30 tubes controlled the card reader and card punch and the remaining tubes helped maintain charges in the condensers. It employed rotating drum memory with each of the two drum memory units able to hold thirty 50-bit numbers (Fig. 5.2).

The ABC was a digital machine, and it was designed for a specific purpose (i.e. solving linear equations) rather than as a general-purpose computer. The working prototype was one of the earliest electronic digital computers.[1] However, the ABC was slow and it required constant operator monitoring.

---

[1] The ABC was ruled to be the first electronic digital computer in the Sperry Rand vs. Honeywell patent case in 1973. However, it was preceded by Zuse's Z3 which appeared in 1941.

G. O'Regan, *Giants of Computing: A Compendium of Select, Pivotal Pioneers*,
DOI 10.1007/978-1-4471-5340-5_5, © Springer-Verlag London 2013

Fig. 5.1 John Atanasoff with
components of ABC

Fig. 5.2 Replica of ABC (Courtesy of Iowa State University)

It used binary mathematics and Boolean logic to solve simultaneous linear
equations. It employed over 270 vacuum tubes for digital computation but had no
central processing unit (CPU) and was not programmable.

It weighed over 300 kg and used 1.6 km of wiring. Data were represented
by 50-bit numbers. It performed 30 additions or subtractions per second. The
memory and arithmetic units could operate and store 60 such numbers at a
time ($60*50 = 3,000$ bits). The arithmetic logic unit was fully electronic and
implemented with vacuum tubes.

The input was in decimal format with standard IBM 80-column punched cards, and the output was decimal via a front panel display. A paper card reader was used as an intermediate storage device to store the results of operations too large to be handled entirely within electronic memory. The ABC pioneered important elements in modern computing including:

- Binary arithmetic and Boolean logic.
- All calculations were performed using electronics rather than mechanical switches.
- Computation and memory were separated.

The ABC was tested and operational by 1942, and its historical significance is that the principles that it employed demonstrated the feasibility of electronic computing. Several of its concepts were used in the ENIAC developed by Mauchly and Eckert.[2]

Atanasoff then commenced a Second World War assignment and worked for the US government in the post-war years. He set up the Ordnance Engineering Corporation in 1952 and sold it to Aerojet Corporation in 1956. He retired in 1960. He received several honours including the US National Medal in Technology which was presented by President Bush in 1990. He died in Maryland aged 91 in 1995.

The ABC was ruled to be the first electronic digital computer in the 1973 *Honeywell vs. Sperry Rand* patent court case in the United States. The court case arose from a patent dispute, and Atanasoff was called as an expert witness in the case. The court ruled that Eckert and Mauchly did not invent the first electronic computer, since the ABC existed as *prior art* at the time of their patent application. It is fundamental in patent law that an invention is novel and that there is no existing prior art. This meant that the Mauchly and Eckert patent application for ENIAC was invalid, and Atanasoff was named by the US court as the inventor of the first digital computer.

5.1   Controversy (Mauchly and Atanasoff)

Mauchly visited Atanasoff on several occasions, and they discussed the implementation of the ABC. Mauchly subsequently designed the ENIAC, EDVAC and UNIVAC. A 1973 legal case ruled that the Atanasoff-Berry Computer (ABC) existed as prior art at the time of the ENIAC patent application. The court ruled that the ABC was the first digital computer and stated that the inventors of ENIAC had derived the subject matter of the electronic digital computer from Atanasoff.

---

[2]John Mauchly and ENIAC are discussed in a later chapter.

Chapter 6
Charles Babbage

Charles Babbage is considered (along with George Boole) to be one of the grandfathers of computing. He made contributions to several areas including mathematics, statistics, astronomy, philosophy, railways and lighthouses. He founded the British Statistical Society and the Royal Astronomical Society (Fig. 6.1).

He was born in Devonshire, England, in 1791 and was the son of a banker. He studied mathematics at Cambridge University in England and was appointed to the Lucasian Chair in Mathematics at Cambridge in 1828.

Babbage was interested in accurate mathematical tables as these are essential for navigation and scientific work. However, there was a high error rate in the existing tables due to human error introduced during calculation. He became interested in the problem of finding a mechanical method to perform the calculations, as this would eliminate errors introduced by human calculation. Pascal invented the *Pascaline* (a simple calculating machine) in the seventeenth century, which was used for performing addition and subtraction. Leibniz subsequently invented a machine called the *Step Reckoner* that could perform addition, subtraction, multiplication and division. Babbage wished to develop a machine to compute polynomial functions.

He designed the Difference Engine (No. 1) in 1821 for the production of mathematical tables. This is essentially a mechanical calculator (analogous to modern electronic calculators), and it was designed to compute polynomial functions. It could also compute logarithmic and trigonometric functions such as sine or cosine (as these may be approximated by polynomials).[1]

---

[1] The power series expansion of the sine function is given by $\mathrm{Sin}(x) = x - x^3/3! + x^5/5! - x^7/7! + \cdots$. The power series expansion for the cosine function is given by $\mathrm{Cos}(x) = 1 - x^2/2! + x^4/4! - x^6/6! + \cdots$. Functions may be approximated by interpolation and the approximation of a function by a polynomial of degree n requires $n+1$ points on the curve for the interpolation. That is, the curve formed by the polynomial of degree n that passes through the $n+1$ points of the function to be approximated is an approximation to the function. The error function also needs to be considered.

G. O'Regan, *Giants of Computing: A Compendium of Select, Pivotal Pioneers*,
DOI 10.1007/978-1-4471-5340-5_6, © Springer-Verlag London 2013

Fig. 6.1 Charles Babbage

The accurate approximation of trigonometric, exponential and logarithmic functions by polynomials depends on the degree of the polynomials, the number of decimal digits that it is being approximated to, and on the error function. A higher-degree polynomial is generally able to approximate the function more accurately.

Babbage produced prototypes for parts of the Difference Engine, but he never actually completed the machine. He also designed the Analytic Engine (the world's first mechanical computer). Its design included a processor, memory and a way to input information and output results.

6.1   Difference Engine

The first working difference engine was built in 1853 by the Swedish engineers George and Edward Scheutz. Their machine was based on Babbage's design, and it was built with funding from the Swedish government. It could compute polynomials of degree 4 on 15-digit numbers. The Science Museum in London has a copy of the 3rd Scheutz Difference Engine on display.

It was the first machine to compute and print mathematical tables mechanically. The machine was accurate, and it showed the potential of mechanical machines as a tool for scientists and engineers.

The difference engine consists of N columns (numbered 1–N). Each column is able to store one decimal number, and the numbers are represented by wheels. The Difference Engine (No. 1) has seven columns with each column containing 20 wheels. Each wheel consists of ten teeth, and these represent the decimal digits. Each column could therefore represent a decimal number with up to 20 digits. The seven columns allowed the representation of polynomials of degree six.

The only operation that the Difference Engine can perform is the addition of the value of column $n + 1$ to column n, and this results in a new value for column n. Column N can only store a constant and column 1 displays the value of the calculation for the current iteration. The machine is programmed prior to execution by setting initial values to each of the columns. Column 1 is set to the value of the polynomial at the start of computation; column 2 is set to a value derived from the first and higher derivatives of the polynomial for the same value of x. Each of the columns from 3 to N is set to a value derived from the $n - 1st$ and higher derivatives of the polynomial.

The Scheutz's difference engine was comprised of shafts, shelves and wheels. The scientist could set numbers on the wheels[2] and turn a crank to start the computation. The decimal numbering system was employed, and there was also a carry mechanism. The scientist could determine the result of the calculation by reading down each shaft. The difference engine was able to print out the answers to the computation.

The machine is unable to perform multiplication or division directly. Once the initial value of the polynomial and its derivatives are calculated for some value of x, the difference engine can calculate any number of nearby values using the numerical method of finite differences. This method replaces computational intensive tasks involving multiplication or division by an equivalent computation which just involves addition or subtraction.

6.2 Finite Differences

A *finite difference* $D[f(x)]$ is a mathematical expression of the form $f(x + h) - f(x)$. If a finite difference is divided by h, then the resulting expression is similar to a differential quotient, except that it is discrete.

Finite differences may be applied to approximate derivatives, and they are often used to find numerical solution to differential equations. They provide a useful way to calculate the degree of the polynomial (and its coefficients) that is used to approximate a given function.

The method of finite differences is used in the production of tables for polynomials using the Difference Engine. Consider the quadratic polynomial $p(x) = 2x^2 + 3x + 1$ and consider the following Table 6.1.

The first difference is computed by subtracting two adjacent entries in the column of $p(x)$. The difference between 15 and 6 is 9, and the difference between 28 and 15 is 13, and so on. The second difference is given by subtracting two adjacent entries

Table 6.1 Finite differences

x	$p(x)$	Diff. 1	Diff. 2
1	6		
2	15	9	
3	28	13	4
4	45	17	4
5	66	21	4

---

[2]Each wheel has ten teeth and represents a decimal digit.

Table 6.2 Finite differences

x	$f(x)$	Diff. 1	Diff. 2
1	6	9	4
2	15	13	4
3	28	17	4
4	45	21	4
5	66	25	4

in the first difference 1 column. That is, the difference between 13 and 9 is 4, and the difference between 17 and 13 is 4, and so on. The entries in the second difference column are the constant 4. In fact, for any n-degree polynomial, the entries in the n-difference column are always a constant.

The Difference Engine performs the computation of the table in a similar manner, although the approach is essentially the reverse of the above. Once the first row of the table has been determined, the rest of the table may be computed by just additions of pairs of cells in the table.

The first row is given by the cells 6, 9 and 4 which allows the rest of the table to be determined. *The numbers in Table 6.2 have been derived by simple calculations from the first row.* The procedure for calculation of the table is as follows:

1. The Difference 2 column is the constant 4.
2. The calculation of the cell in row i for the Difference 1 column is given by Diff. $1(i-1) + $ Diff. 2 $(i-1)$.
3. The calculation of the cell in row i for the function column is given by $f(i-1) + $ Diff. 1 $(i-1)$.

In other words, *to calculate the value of a particular cell, all that is required is to add the value in the cell immediately above it to the value of the cell immediately to its right.* Therefore, in the second row, the calculations $6+9$ yields 15, and $9+4$ yields 13, and since the last column is always the constant 4, it is just repeated. Therefore, the second row is 15, 13 and 4 and $f(2) = 15$. Similarly, the third row yields $15+13 = 28$, $13+4 = 17$ and so on. This is the underlying procedure of the Difference Engine.

The initial problem is to compute the first row which allows the other rows to be computed. Its computation is more difficult for complex polynomials. *The second problem is to find a suitable polynomial to represent the function, and this may be done by interpolation.* However, once these problems are solved the engine produces pages and columns full of data.

Babbage received £17 K of taxpayer funds to build the Difference Engine. However, due to personal reasons he only produced prototypes of the intended machine. The prototypes built by his engineer Joseph Clement were limited to the computation of quadratic polynomials of six digit numbers. Babbage intended that the machine would operate on 6th degree polynomials of 20 digits. The British government cancelled the project in 1842.

He designed an improved difference engine (No. 2) in 1849 (Fig. 6.2). It could operate on 7th-order differences (i.e. polynomials of order 7) and 31-digit numbers. The machine consisted of eight columns with each column consisting of 31 wheels.

Fig. 6.2   Difference Engine No. 2 (Photo Public Domain)

However, it was over 150 years later before it was built (in 1991) to mark the two hundredth anniversary of his birth. The Science Museum in London also built the printer that Babbage designed, and both the machine and the printer worked correctly according to Babbage's design (after a little debugging).

6.3   Analytic Engine

The Difference Engine was designed to produce mathematical tables and required human intervention to perform the calculation. Babbage recognized its limitations and proposed a revolutionary solution. His plan was to construct a new machine that would be capable of executing all tasks that may be expressed in algebraic notation. The Analytic Engine envisioned by Babbage consisted of two parts (Table 6.3):

Babbage intended that the operation of the Analytic Engine would be analogous to the operation of the *Jacquard loom*.[3] The latter is capable of weaving (i.e. executing on the loom) a design pattern that has been prepared by a team of skilled

---

[3]The Jacquard loom was invented by Joseph Jacquard in 1801. It is a mechanical loom which used the holes in the punched cards to control the weaving of patterns in a fabric. The use of punched cards allowed complex designs to be woven from the pattern defined on the punched card. Each punched card corresponds to one row of the design and the cards were appropriately ordered. It was very easy to change the pattern of the fabric being weaved on the loom, as this simply involved changing the cards.

Table 6.3 Analytic Engine

Part	Function
Store	This contains the variables to be operated upon as well as all those quantities which have arisen from the result of intermediate operations
Mill	The mill is essentially the processor of the machine into which the quantities about to be operated upon are brought

artists. The design pattern is represented by punching holes on a set of cards, and each card represents a row in the design. The cards are then ordered and placed in the loom, and the exact pattern is produced by the loom.

The Jacquard loom was the first machine to use punched cards to control a sequence of operations. It did not perform computation, but it was able to change the pattern of what was being weaved by changing cards. This gave Babbage the idea to use punched cards to store programs to perform the analysis and computation in the Analytic Engine.

The use of the punched cards in the Analytic Engine allowed the formulae to be manipulated in a manner dictated by the programmer. The cards commanded the analytic engine to perform various operations and to return a result. Babbage distinguished between two types of punched cards:

– *Operation cards*
– *Variable cards*

Operation cards are used to define the operations to be performed, whereas the variable cards define the variables or data that the operations are performed upon. His planned use of punched cards to store programs in the Analytic Engine is similar to the idea of a stored computer program in von Neumann architecture. However, Babbage's idea of using punched cards to represent machine instructions and data was over 100 years before digital computers. *Babbage's Analytic Engine is therefore an important milestone in the history of computing*.

The Analytic Engine was designed in 1834 as the world's first mechanical computer [Bab:42]. It included a processor, memory and a way to input information and output results. However, the machine was never built as Babbage was unable to receive funding from the British government.

Babbage intended that the program be stored on read-only memory using punched cards, and that the input and output would be carried out using punched cards. He intended that the machine would be able to store numbers and intermediate results in memory that could then be processed. There would be several punched card readers in the machine for programs and data. He envisioned that the machine would be able to perform conditional jumps as well as parallel processing where several calculations could be performed at once.

Chapter 7
John Backus

John Backus was an American computer scientist and language designer. He made important contributions to the development of the FORTRAN programming language, and he also developed Backus-Naur Form (BNF), which is a widely used notation for expressing the syntax of a programming language.

He was born in Philadelphia in 1924 and grew up in Delaware. He served in the US army and briefly considered a career in medicine. He discovered that he had an aptitude for mathematics shortly after the war, while he was studying to become a radio technician. He pursued a postgraduate degree at Columbia University in New York and was awarded a Master's in mathematics in 1949 (Fig. 7.1).

He joined IBM in 1950, and his initial work was on scientific computation using the IBM Special Sequence Electronic Calculator (SSEC) to compute lunar orbital positions. He developed the first high-level language for an IBM computer in 1953, and this language was called "Speedcode" (or "Speedcoding"), and it was used on the IBM 701 computer. This interpreted language was designed to support the computation of floating point numbers to calculate astronomical positions. It included instructions for common mathematical and scientific functions such as logarithms, exponentials and trigonometric functions.

It was not very efficient in program execution, as it took 10–20 times longer to execute a program through the Speedcode interpreter than with the execution of the equivalent machine code. However, its advantage [Bac:53] was that it led to significant programmer productivity improvements, leading to major savings as programming, testing and debugging were completed faster.

Programming in the early 1950s was done using binary machine code, which was slow, frustrating and error prone. Backus believed that there must be an easier way to do programming, where the instructions for a program could be written in a language resembling English and then translated by the computer into the equivalent machine code.

This led to his interest in the development of an efficient high-level programming language that could be used for scientific applications such as weather forecasting and aircraft design. He discussed his ideas on the development of a practical high-level programming system with Dr. Cuthbert Hurd, the head of the Applied

G. O'Regan, *Giants of Computing: A Compendium of Select, Pivotal Pioneers*,
DOI 10.1007/978-1-4471-5340-5_7, © Springer-Verlag London 2013

Fig. 7.1 John Backus

Science Department at IBM, and he was appointed to lead a team that developed
the FORTRAN programming language. The language was designed for scientific
and technical computation, and a compiler for the language for the IBM 704 was
available in 1957.

FORTRAN led to a massive increase in programmer productivity, and what had
previously taken 1,000 machine instructions could now be written in 47 statements.
Engineers and scientists could now learn how to do their own programming, and
compilers for the language were developed for other IBM machines and later for
other computers. FORTRAN remains the most widely used programming language
for scientific applications.

Backus served on the international program committee which developed the
Algol-58 and Algol-60 programming languages. Algol became a popular language
for the expression of algorithms, and Backus developed a formal technique known
as *Backus Normal Form* or *Backus-Naur*[1] *Form* (BNF) to express the syntax of a
programming language. The syntax (or grammar) of a language is used to determine
if a program is syntactically correct.

He was named an IBM fellow in 1963 in recognition of his contribution to
language design, and he was awarded the *W.W. McDowell Award* for outstanding
contributions to the computing field from the *Institute of Electrical and Electronic
Engineers* (IEEE) in 1967. He received the *ACM Turing Award* in 1977 for his
contributions to the design and development of high-level programming languages
(especially FORTRAN).

The title of his Turing Award lecture was "Can programming be liberated from
the von Neumann style" [Bac:78]. He noted that most programming languages
are based on the von Neumann model of computing, and that such programs are
difficult to reason about. He argued that each successive programming language
includes the features of its predecessors plus a few more, but that few languages

[1]Peter Naur is a Danish computer scientist who contributed to the development of the Algol
programming language and was editor of the report describing the language.

make programming significantly cheaper or more reliable. He argued that there was a need for a powerful methodology to reason about programs, and that conventional programming languages create unnecessary confusion in thinking about programs.

He noted that the essential architecture of a computer system remains von Neumann. It consists of a central processing unit (CPU), a store and a connecting tube that can transmit a single word between them. The task of the program is to change the contents of the store in a major way, and this involves pumping words back and forth through the *von Neumann bottleneck*. The assignment statement is the von Neumann bottleneck of programming languages, as it makes the programmer think of words at a time in the same way that the computer's bottleneck does.

His Turing award paper argues for a function-level-programming paradigm in which programs are treated as mathematical objects. He later designed FP to be a language that would support the function-level programming style, where a function-level program is variable free, and a program is built directly from given programs using functions. This enables the set of programs to form a mathematical space with nice algebraic properties. Next, we discuss FORTRAN and BNF in more detail.

7.1 FORTRAN

FORTRAN (FORmula TRANslator) was the first high-level programming language to be implemented. It was developed by John Backus at IBM in the mid-1950s, and the first compiler was available in 1957. The language includes named variables, complex expressions and subprograms. It was designed for scientific and engineering applications, and it remains the most important programming language for these domains. The main statements of the language include:

- Assignment statements (using the = symbol)
- IF statements
- Goto statements
- DO Loops

FORTRAN II was developed in 1958, and it introduced subprograms and functions to support procedural programming. Each procedure (or subroutine) contains computational steps to be carried out and may be called at any point during program execution. This could include calls to other procedures. However, recursion (i.e. a procedure call to itself) was not allowed until FORTRAN 90. FORTRAN 2003 provides support for object-oriented programming.

The basic types supported in FORTRAN include Boolean, Integer and Real. Support for double precision and complex numbers was added later. The language included relational operators for equality (.EQ.), less than (.LT.) and so on. It was good at handling numbers and computation, which was especially useful for scientific and engineering problems. The following code (written in Fortran 77) gives a flavour of the language.

```
        PROGRAM HELLOWORLD
C       FORTRAN 77 SOURCE CODE COMMENTS FOR HELLOWORLD
        PRINT '(A)', 'HELLO WORLD'
        STOP
        END
```

FORTRAN remains a popular programming language for scientific applications such as climate modelling and simulations of the solar system.

7.2   Backus-Naur Form

There are two key parts to any programming language, namely, its syntax and semantics. The syntax is the grammar of the language, and the programs need to be grammatically correct prior to execution. The semantics of the language is deeper and determines the meaning of what has been written by the programmer.

The theory of the syntax of programming languages is well established, and Backus-Naur Form (BNF) is employed to specify the grammar of context-free languages. The grammar of a language may be input into a parser, which determines whether the program is syntactically correct or not. A BNF specification consists of a set of rules such as:

<symbol> ::= <expression with symbols>

where <symbol> is a *nonterminal* and the expression consists of sequences of symbols and/or sequences separated by the vertical bar "|" which indicates a choice. Symbols that never appear on the left side of a rule are called *terminals*, and those on the right-hand side consist of terminals and nonterminals. A partial definition of the syntax of statements in a programming language is given by:

<loop statement> ::= <while loop>|<for loop>
<while loop> ::= while (<condition>) <statement>
<for loop> ::= for (<expression>) <statement>
<statement> ::= <assignment statement>|<loop statement>
<assignment statement> ::= <variable> := <expression>

This example includes several nonterminals (e.g. <loop statement>, <while loop>) and terminals (e.g. "while", "for"). The grammar of a language is translated by a parser into a parse table. Each type of grammar has its own parsing algorithm to determine whether a particular program is valid with respect to the grammar.

Chapter 8
Gordon Bell

Gordon Bell was vice president of Research and Development at Digital Equipment Corporation (DEC) from 1960 to 1983. He was the architect of various PDP computers and led the development of the VAX series of computers. He has been involved in the design of around 30 microprocessors (Fig. 8.1).

He was born in Missouri in 1934 and obtained BS and MS degrees in electrical engineering from MIT in 1956 and 1957, respectively. He then went to the University of New South Wales on a Fulbright scholarship, where he taught computer design and programmed the UTECOM. He returned to the United States and joined the Speech Computation Laboratory at MIT.

Olsen and Anderson founded Digital Equipment Corporation (DEC) in 1959, and Bell knew both of them from his time at MIT. He joined the newly formed company as one of its earliest employees in 1960 and was involved in the development of the Programmed Data Processor (PDP) family of minicomputers. He designed the multiplier/divider unit and the interrupt system for the PDP-1 computer. The PDP-1 computer was built upon the work done at the MIT Lincoln Laboratory.

He was the architect for the PDP-4 and PDP-6 computers, and he took leave of absence from DEC from 1966 to 1972 to become professor of computer science and electrical engineering at Carnegie Mellon University. He returned to DEC in 1972 as vice president of engineering for overseeing the development of the 32-bit VAX series of computers. These were a highly successful product line for the company.

He resigned from DEC in 1983 and became one of the founders of Encore Computer. This Massachusetts-based company was a pioneer in the parallel computing market, and its goal was to build massively parallel machine. He served as assistant director with the National Science Foundation's Computing and Information Science and Engineering Directorate from 1986 to 1987.

He established the *ACM Gordon Bell Prize* 1987 to encourage developments in parallel processing. This award recognizes outstanding contributions to high-performance computing.

G. O'Regan, *Giants of Computing: A Compendium of Select, Pivotal Pioneers*,
DOI 10.1007/978-1-4471-5340-5_8, © Springer-Verlag London 2013

Fig. 8.1 Gordon Bell
(Courtesy of Queensland
University of Technology)

He assisted many small start-up companies including Stardent Computer and Ardent Corporation. He was a co-founder and director of the Bell Mason Group, which supplied expert systems to analyze the strengths and weaknesses of new businesses. This was especially useful to start-up companies and investors.

He advised Microsoft in the early 1990s in its goals to set up a research group and became involved full time with Microsoft Research in the United States from the mid-1990s. His research interests are in scalable systems, high-performance computing and trends in supercomputing. He remains a principal researcher at Microsoft Research and continues to maintain an interest in start-up companies.

He was a co-founder with Ken Olsen of the *Computer History Museum* in California in 1999. He had previously co-founded the Computer Museum in Boston in 1979, and this collection was moved to the Computer History Museum. He is the author of several books on computer design and entrepreneurship. These include *High Tech Ventures: The Guide for Entrepreneurial Success*, *Computer Engineering* and *Total Recall: How the E-memory Revolution Will Change Everything*.

He has received several awards in recognition of his contributions to the computing field. He received the *National Medal in Technology* in 1991 for his industrial achievement in the computer field from President George Bush. He received the *IEEE John von Neumann medal* in 1992. He became a fellow of the Computer History Museum in 2003.

8.1   Bell's Law of Computer Classes

Bell's Law of Computer Classes was formulated by Bell in 1972 and describes how types of computing systems (*computer classes*) form, evolve and eventually die out. Bell argues that new computer classes are formed roughly every decade, and the new class creates new applications, resulting in new markets and new industries.

The new class is enabled by advances in technology such as semiconductors, networks, storage and interfaces. This allows a new lower price computer class to be created that serves a new need in the market place. It may be smaller device based on a new programming platform and network, and it leads to the establishment of a new industry. For example, mainframe computers dominated the 1960s, microcomputers were popular in the 1970s, personal computers and local area networks were popular in the 1980s, personal computers and web browsers and the Internet became dominant in the 1990s and web services became important from 2002.

8.2 PDP and VAX Series of Minicomputers

Bell was the architect of several members of the PDP family at DEC (including the PDP 06). He initially worked on the interrupt system of the PDP-1 machine, and the PDP-1 was the first of a long line of DEC computers that focused on interaction with the user and on affordability. There is a restored version of the PDP-1 computer at the Computer History Museum in California.

The PDP-8 minicomputer was a 12-bit machine with a small instruction set, and it was released in 1965. It was a major commercial success for DEC with many sold to schools and universities. The PDP-11 was a highly successful series of 16-bit minicomputer, and it remained a popular product for over 20 years from the 1970s to the 1990s.

DEC's next generation of computers was the 32-bit VAX series of computers, and these were introduced following the return of Bell as the VP of engineering in 1972. The VAX series were derived from the PDP-11, and it was the first widely used 32-bit minicomputer. The VAX 11/780 was released in 1978, and it was a major success for the company. It was one of the most successful families of computers of all time.

The VAX product line was a competitor to the IBM System/370 series of computers, and it used the Virtual Operating System known as VMS.

8.3 Computer History Museum

Bell co-founded (with Ken Olsen) the Computer History Museum in Mountain View, California, in 1999. He had previously co-founded the Computer Museum in Boston with his wife in 1979, and DEC had supplied many historical artefacts for this collection. The initial collection of the Computer History Museum came from the Computer Museum, and the museum has a rich collection of exhibits related to the history of computing.

The collection includes both online exhibits as well as physical exhibits. It covers the first 25 years of microprocessors, computer chess, the IBM Stretch computer, a hall of fellows who have made outstanding contributions to the computing field, a history of the Internet from 1962 to 1992, a computer history timeline of important events in the history of computing, the restored PDP-1 computer and information on Babbage's Difference Engine and a replica of Babbage's Difference Engine No. 2.

The museum organizes regular events and publishes the *Core Magazine* which presents articles on the historical roots of computer technology. The museum has a catalogue search facility which is a useful resource for students and researchers. Bell became a fellow of the museum in 2003.

Chapter 9
Tim Berners-Lee

Tim Berners-Lee is a British computer scientist and the inventor of the World Wide Web. He was born in London in 1955 and obtained a degree in physics in 1976 from Oxford University. Both his parents had been involved in the programming of the Ferranti Mark I computer in the 1950s (Fig. 9.1).

He studied physics at Oxford University from 1973 to 1976 and obtained his degree in 1976. He then worked at Plessey[1] and joined D.G. Nash[2] in 1978. He then became an independent consultant and went to CERN in the 1980s for a short-term contract programming assignment. CERN is a key European centre for research in the nuclear field, and it employs several thousand physicists and scientists.

One of the problems that CERN faced in the 1980s was keeping track of people, computers, documents and databases. The centre had many visiting scientists who spent several months there, as well as a large pool of permanent staff. There was no effective way to share information among scientists at that time, and this motivated Berners-Lee to investigate an information management tool to support physicists in their work.

A visiting scientist might need to obtain information or data from a CERN computer or to make their research results available to other colleagues at CERN or to colleagues throughout the world. Berners-Lee developed a prototype information management program called "ENQUIRE"[3] to assist with information sharing among researchers and in keeping track of the work of visiting scientists.

He returned to CERN in 1984 to work on other projects and devoted part of his free time to consider solutions to the information-sharing problem. His solution to this problem was the invention of the *World Wide Web* in 1990.

---

[1]Plessey was a large British telecommunications company based in Poole in England. It was taken over by GEC and Siemens in 1989.

[2]D.G. Nash was a small company run by Denis Nash and John Poole.

[3]The name "ENQUIRE" came from a popular Victorian book called *Enquire Within Upon Everything* which was originally published in 1850.

Fig. 9.1 Tim Berners-Lee
(Courtesy of Uldis Bojārs)

He built on several existing inventions such as the *Internet*, *hypertext* and the *mouse*. Ted Nelson invented hypertext in the 1960s, and it allows links to be present in text. For example, a document such as a book contains a table of contents, an index and a bibliography. These are all links to material that is either within the book itself or external to the book. The reader of a book is able to follow the link to obtain the internal or external information. Doug Engelbart invented the mouse in the 1960s, and it allows the cursor to be steered around the screen.

The major leap that Berners-Lee made was essentially a marriage of the Internet, hypertext and the mouse into what has become the World Wide Web. His vision [BL:00] and its subsequent realization benefited CERN and the wider world.

> Suppose that all information stored on computers everywhere were linked. Program computer to create a space where everything could be linked to everything.

The World Wide Web creates a space in which users can access information easily in any part of the world. This is done using only a web browser and simple web addresses. The user can then click on hyperlinks on web pages to access further relevant information that may be on an entirely different continent. Berners-Lee is the director of the World Wide Web Consortium, and this MIT-based organization sets the software standards for the web.

The invention of the World Wide Web was a revolutionary milestone in computing. It transformed the use of the Internet from mainly academic use to where it is now an integral part of peoples' lives. Users may now *surf the web*: i.e. hyperlink among the millions of computers in the world and obtain information easily. It is revolutionary in that:

- No single organization is controlling the web.
- No single computer is controlling the web.
- Millions of computers are interconnected.
- It is an enormous market place of billions of users.
- The web is not located in one physical location.
- The web is a space and not a physical thing.

Table 9.1 Features of the World Wide Web

Feature	Description
URL	Universal resource identifier (later renamed to universal resource locator (URL)) provides a unique address code for each web page
HTML	Hypertext markup language (HTML) is used for designing the layout of web pages
HTTP	The hypertext transfer protocol (HTTP) allows a new web page to be accessed from the current page
Browser	A browser is a client program that allows a user to interact with the pages and information on the World Wide Web

He created a system that gives every web page a standard address called the universal resource locator (URL). Each page is accessible via the hypertext transfer protocol (HTTP), and the page is formatted with the hypertext markup language (HTML). Each page is visible using a web browser. The key features of Berners-Lee invention are shown in Table 9.1.

Berners-Lee invented the well-known terms such as URL, HTML and World Wide Web. He wrote the first browser program that allowed users to access web pages throughout the world. Browsers are used to connect to remote computers over the Internet and to request, retrieve and display the web pages on the local machine. The first website was at CERN (info.cern.ch).

The early browsers included Gopher developed at the University of Minnesota and Mosaic developed at the University of Illinois. Netscape replaced these and dominated the browser market until Microsoft developed Internet Explorer. The development of graphical browsers led to the commercialization of the World Wide Web.

9.1 Applications of the World Wide Web

Berners-Lee realized that the World Wide Web offered the potential to conduct business in cyberspace, rather than the traditional way of buyers and sellers coming together in the market place.

> Anyone can trade with anyone else except that they do not have to go to the market square to do so.

The growth of the World Wide Web has been phenomenal, and exponential growth rate curves became a feature of newly formed Internet companies and their business plans. The World Wide Web has been applied to many areas including:

- Travel industry (booking flights, train tickets and hotels)
- E-Marketing
- Online shopping (e.g. www.amazon.com)
- Recruitment services (such as www.jobserve.com)

Table 9.2 Characteristics of e-commerce

Feature	Description
Catalogue of products	The catalogue of products details the products available for sale and their prices
Well-designed and easy to use	The usability of a website is a key concern
Shopping carts	This is analogous to shopping carts in a supermarket
Security	Security of credit card information is a key concern for users of an e-commerce site
Payments	There is a check-out facility to arrange for the purchase of the goods
Order fulfilment	Once payment has been received the products must be delivered to the customer

- Internet banking
- Newspapers and news channels
- Social media (e.g. Facebook)

The prediction in the early days was that the new web-based economy would replace traditional bricks and mortar companies and that most business would be conducted over the web. The size of the new web economy was estimated to be in trillions of US dollars.

New companies were formed to exploit the opportunities of the web, and existing companies developed e-business and e-commerce strategies. The new business models included business-to-business (B2B) and business-to-consumer (B2C). E-commerce websites have the characteristics listed in Table 9.2.

Berners-Lee has received many awards in recognition of his achievements. He was knighted by Queen Elizabeth in 2004 and has received several honorary doctorates. *Time* magazine named him in 1999 as one of the 100 most important people of the twentieth century.

Chapter 10
Dines Bjørner

Dines Bjørner is a Danish computer scientist who has made important contributions to software engineering and formal methods (Fig. 10.1). He developed the Vienna Development Method (VDM) with Cliff Jones at the IBM laboratory in Vienna. VDM was one of the earliest formal methods, and it is used to increase confidence in the correctness of software in academia and industry. He was also involved in the development of the RAISE (Rigorous Approach to Industrial Software Engineering) method and set of tools.

He was born in Odense, Denmark, in 1937, and he studied at the Technical University of Denmark. He obtained an MSc degree in Electronics Engineering in 1962 and earned a Ph.D. in computer science from the university in 1969. He joined IBM in 1962 and worked on hardware design on the IBM 1070, IBM 1800 and IBM 1130 in California and at IBM's research facility in Switzerland. He transferred to the IBM Vienna Laboratory, which was directed by Heinz Zemanek, and he worked with Cliff Jones, Peter Lucas and others on the semantics of the PL/1 programming language. The group developed the Vienna Development Method to assist in the definition of the formal semantics of the language, and he would later apply VDM to the specification of software systems and to practical industrial problems. He resigned from IBM in 1975 and returned to Denmark.

He became professor at the Technical University of Denmark from 1976 to 2007. He was responsible for setting up the International Institute for Software Technology (UNU-IIST) at the United Nations University in Macau in 1992, and he was the first director of the institute.

He was the co-founder of VDM-Europe in 1987 (which became FME-Europe in 1991). He is the author of several books and over a hundred papers. His books include a three-volume work on software engineering as well as a book with Cliff Jones on the Vienna Development Method. He has received several awards for his contributions to the computer field.

G. O'Regan, *Giants of Computing: A Compendium of Select, Pivotal Pioneers*,
DOI 10.1007/978-1-4471-5340-5_10, © Springer-Verlag London 2013

Fig. 10.1 Bjørner at ICTAC
2007

10.1   Formal Methods

The term *formal methods* refers to various mathematical techniques used in the
software field for the formal specification and development of software. A formal
method includes a formal specification language and a collection of tools to support
the syntax checking of the specification, as well as the proof of properties of
the specification. Abstraction is employed and this allows questions about what
the system does to be answered independently of the implementation. The use of
mathematical notation helps to avoid the ambiguity inherent in natural language.

The specification is written in a mathematical language, and the implementation
is derived from the specification via stepwise refinement. The refinement step makes
the specification more concrete and closer to the actual implementation. There is
an associated proof obligation that the refinement is valid, and that the concrete
state preserves the properties of the more abstract state. Thus, assuming that the
original specification is correct and the proofs of correctness of each refinement
step are valid, then there is a very high degree of confidence in the correctness of
the implemented software.

Requirements are the foundation from which the system is built, and irrespective
of the best design and development practices, the product will be incorrect if the
requirements are incorrect. Formal methods may be employed to model the requirements, and model exploration yields further desirable or undesirable properties. The
ability to prove that certain properties are true of the specification is very valuable,
especially in safety critical and security critical applications. These properties are
logical consequences of the definition of the requirements, and, if appropriate, the
requirements may need to be amended. Thus, formal methods may be employed for
requirements validation and in a sense to *debug the requirements*.

The use of formal methods generally leads to more robust software and to
increased confidence in its correctness. They have been applied to a diverse range
of applications, including the security critical field, the safety critical field, the
railway sector, microprocessor verification, the specification of standards and the
specification and verification of programs.

10.2 Vienna Development Method

VDM was developed at the IBM research laboratory in Vienna in the late 1960s. The group was working on the definition of the semantics of the PL/1 programming language and employed a specification language termed the Vienna Definition Language (VDL). They initially employed an operational semantic approach; i.e. the semantics of a language are determined in terms of a hypothetical machine which interprets the programs of that language [BjJ:78]. Later work led to the Vienna Development Method (VDM) with its specification language, Meta IV. VDM was employed to give the denotational semantics of PL/1; i.e. a mathematical object (set, function, etc.) is associated with each phrase of the language [BjJ:82]. The mathematical object is the *denotation* of the phrase. The initial application of VDM was to programming language semantics. Today, it is employed to formally specify software and includes a formal development method.

The Vienna group was broken up in the mid-1970s, and this led to the formation of several schools of VDM in diverse locations. The "Danish school" was led by Dines Bjørner, the English school by Cliff Jones and the Polish school by Andrez Blikle [ORg:06]. Further work on VDM continued in the 1980s, and standards (VDM-SL) appeared in the 1990s.

VDM is a *model-oriented approach*, and this means that an explicit model of the state is given, and operations are defined in terms of this state. The model acts as a representation of the proposed system, and it is explored to determine its suitability. This involves asking questions, and the effectiveness of the model is judged by its ability to answer the questions. The modelling involves using discrete mathematics: e.g. set theory, sequences, functions and relations.

Operations may act on the system state, taking inputs and producing outputs and a new system state. Operations are defined in a precondition and postcondition style. Each operation has an associated proof obligation to ensure that if the precondition is true, then the operation preserves the system invariant. The initial state itself is, of course, required to satisfy the system invariant.

VDM uses keywords to distinguish different parts of the specification, e.g. preconditions and postconditions, which are introduced by the keywords *pre* and *post*, respectively. VDM employs postconditions to stipulate the effect of the operation on the state. The previous state is then distinguished by employing *hooked variables*, e.g. $\overset{\leftharpoonup}{v}$, and the postcondition specifies the new state (*defined by a logical predicate relating the pre-state to the post-state*).

VDM is more than its specification language Meta IV (called VDM-SL in the standardization of VDM). It is a development method, with rules to verify the steps of development. These rules enable the executable specification, i.e. the detailed code, to be obtained from the initial specification via refinement steps. Thus, we have a sequence $S = S_0, S_1, \ldots, S_n = E$ of specifications, where S is the initial specification and E is the final (executable) specification.

$$S = S_0 \sqsubseteq S_1 \sqsubseteq S_2 \sqsubseteq \ldots \sqsubseteq S_n = E$$

Retrieval functions enable a return from a more concrete specification to the more abstract specification. The initial specification consists of an initial state, a system state and a set of operations. The system state is a particular domain, where a domain is built out of primitive domains such as the set of natural numbers or constructed from primitive domains using domain constructors such as Cartesian product and disjoint union. A domain-invariant predicate may further constrain the domain, and a *type* in VDM reflects a domain obtained in this way. Thus, a type in VDM is more specific than the signature of the type and thus represents values in the domain defined by the signature, which satisfy the domain invariant.

VDM specifications are structured into modules, with a module containing the module name, parameters, types, operations, etc. Partial functions arise naturally in computer science. The problem is that many functions, especially recursively defined functions, can be undefined or fail to terminate for some arguments in their domain. VDM addresses partial functions by employing nonstandard logical operators, namely, the logic of partial functions (LPFs) which can deal with undefined operands. A VDM specification consists of:

- Type definitions
- State definitions
- Invariant for the system
- Definition of the operations of the system

There are many examples of VDM specifications [InA:91].

10.3   RAISE

RAISE (Rigorous Approach to Industrial Software Engineering) was a European project led by Dines Bjørner in the 1990s. RAISE is a formal method designed to assist industry in the development of high-quality software.

The RAISE method is based on the stepwise refinement paradigm as in VDM and Z. RAISE consists of a set of tools (the RAISE tools) and the RAISE specification language (RSL). The RAISE tools support editing of specifications, proof of properties, translators from specifications into C++ and Ada and a document formatter. It supports model-oriented and algebraic specifications.

RAISE has been used on several industrial projects. It is promoted by the United Nations University in Macau (UNU-IIST).

Chapter 11
George Boole

Boole was born in Lincoln, England, in 1815. His father (a cobbler who was interested in mathematics and optical instruments) taught him mathematics and showed him how to make optical instruments. George Boole inherited his father's interest in knowledge and was self-taught in mathematics and Greek. He taught in various schools near Lincoln and developed his mathematical knowledge by working his way through Newton's Principia, as well as applying himself to the work of mathematicians such as Laplace and Lagrange (Fig. 11.1).

He published regular papers from his early twenties, and these included contributions to probability theory, differential equations and finite differences. He developed Boolean algebra which is the foundation for modern computing, and he is considered (along with Babbage) to be one of the grandfathers of computing. *His work was theoretical, and he never actually built a computer or calculating machine. However, Boole's symbolic logic was the perfect mathematical model for switching theory and for the design of digital circuits.*

Boole became interested in formulating a calculus of reasoning, and he published a pamphlet titled "Mathematical Analysis of Logic" in 1847 [Boo:48]. This article developed novel ideas on a logical method, and he argued that logic should be considered as a separate branch of mathematics, rather than a part of philosophy. He argued that there are mathematical laws to express the operation of reasoning in the human mind, and he showed how Aristotle's syllogistic logic could be reduced to a set of algebraic equations. He corresponded regularly on logic with Augustus De Morgan.[1]

---

[1] De Morgan was a nineteenth-century British mathematician based at University College London. De Morgan's laws in Set Theory and Logic state that $(A \cup B) = A^c \cap B^c$ and $\neg (A \vee B) = \neg A \wedge \neg B$.

G. O'Regan, *Giants of Computing: A Compendium of Select, Pivotal Pioneers*, DOI 10.1007/978-1-4471-5340-5_11, © Springer-Verlag London 2013

Fig. 11.1 George Boole

Boole had no formal university qualification, and he had difficulty in obtaining a university position. However, his publications were recognized as excellent,[2] and he was awarded the position as the first professor of mathematics at the newly founded Queens College Cork[3] in 1849.

His paper on logic introduced two quantities "0" and "1". He used the quantity 1 to represent the universe of thinkable objects (i.e. the universal set), and the quantity 0 represents the absence of any objects (i.e. the empty set). He then employed symbols such as x, y and z to represent collections or classes of objects given by the meaning attached to adjectives and nouns. Next, he introduced three operators $(+, -$ and $\times)$ that combined classes of objects.

The expression xy (i.e. x multiplied by y or $x \times y$) combines the two classes x, y to form the new class xy (i.e. the class whose objects satisfy the two meanings represented by the classes x and y). Similarly, the expression $x + y$ combines the two classes x, y to form the new class $x + y$ (that satisfies either the meaning represented by class x or class y). The expression $x - y$ combines the two classes x, y to form the new class $x - y$. This represents the class that satisfies the meaning represented by class x but not class y. The expression $(1 - x)$ represents objects that do not have the attribute that represents class x.

Thus, if $x =$ black and $y =$ sheep, then xy represents the class of black sheep. Similarly, $(1 - x)$ would represent the class obtained by the operation of selecting all things in the world except black things; $x (1 - y)$ represents the class of all things that are black but not sheep, and $(1 - x) (1 - y)$ would give us all things that are neither sheep nor black.

---

[2]Boole was awarded the Royal Medal from the Royal Society of London in 1844 in recognition of his publications. The Irish mathematician Sir Rowan Hamilton (who invented quaternions) was another famous recipient of this prize.

[3]Queens College Cork is now called University College Cork (UCC) and has about 18,000 students. It is located in Cork City in the south of Ireland.

He showed that these symbols obeyed a rich collection of algebraic laws and could be added, multiplied, etc., in a manner that is similar to real numbers. These symbols may be used to reduce propositions to equations, and algebraic rules may be employed to solve the equations. The rules include:

1.	$x + 0 = x$	(Additive identity)
2.	$x + (y + z) = (x + y) + z$	(Associative)
3.	$x + y = y + x$	(Commutative)
4.	$x + (1 - x) = 1$	
5.	$x\,1 = x$	(Multiplicative identity)
6.	$x\,0 = 0$	
7.	$x + 1 = 1$	
8.	$xy = yx$	(Commutativity)
9.	$x(yz) = (xy)z$	(Associativity)
10.	$x(y + z) = xy + xz$	(Distributive)
11.	$x(y - z) = xy - xz$	(Distributive)
12.	$x^2 = x$	(Idempotent)

These operations are similar to the modern laws of set theory with the set union operation represented by "+", and the set intersection operation is represented by multiplication. The universal set is represented by "1" and the empty by "0". The associative and distributive laws hold. Finally, the set complement operation is given by $(1 - x)$.

He applied the symbols to encode Aristotle's syllogistic logic, and he showed how the syllogisms could be reduced to equations. This allowed conclusions to be derived from premises by eliminating the middle term in the syllogism. He refined his ideas on logic further in his book *An Investigation of the Laws of Thought* [Boo:58]. This book aimed to identify the fundamental laws underlying reasoning in the human mind and to give expression to these laws in the symbolic language of a calculus.

He considered the equation $x^2 = x$ to be a fundamental law of thought. It allows the principle of contradiction to be expressed (i.e. for an entity to possess an attribute and at the same time not to possess it):

$$x^2 = x$$

$$\Rightarrow x - x^2 = 0$$

$$\Rightarrow x(1 - x) = 0$$

For example, if x represents the class of horses, then $(1 - x)$ represents the class of "not-horses". The product of two classes represents a class whose members are common to both classes. Hence, $x\,(1 - x)$ represents the class whose members are at once both horses and "not-horses", and the equation $x\,(1 - x) = 0$ expresses that fact that there is no such class. That is, it is the empty set.

Boole contributed to other areas in mathematics including differential equations and finite differences[4] and to the development of probability theory. He married Mary Everest in 1855, and they lived in Lichfield Cottage in Ballintemple, Cork. She was a niece of Sir George Everest, after whom the world's highest mountain is named. Boole died from fever at the early age of 49 in 1864.

Queens College Cork honoured his memory by installing a stained glass window in the *Aula Maxima* of the college. It showed Boole writing at a table with Aristotle and Plato in the background. The annual Boole Prize is awarded by the mathematics department at University College Cork, and a UCC mathematician Des McHale has written an interesting biography of Boole [McH:85].

Boole's work on logic appeared to have no practical use. However, Claude Shannon's 1937 Master's thesis showed that it provided the perfect mathematical model for switching theory and for the design of digital circuits.

The use of the properties of electrical switches to process logic is the basic concept that underlies all modern electronic digital computers. Digital computers today use the binary digits 0 and 1, and Boolean logical operations may be implemented by electronic AND, OR and NOT gates. More complex circuits (e.g. arithmetic) may be designed from these fundamental building blocks.

11.1   Modern Boolean Algebra

Boolean algebra deals two values 1 and 0 (i.e. true and false). It deals with propositions that are either true or false and employs logical operators such as "and, $\wedge$", "or, $\vee$" and "not, $\neg$". Other logical operators are described in [ORg:06]. The proposition "$2 + 2 = 4$" is true, whereas the proposition "$2 * 5 = 11$" is false. Variables (e.g. A, B) are used to stand for propositions and may be combined using logical operators to form new propositions. Several well-known properties of Boolean algebra are detailed in Table 11.1.

The conjunction of any operand B with the Boolean value "True" yields the proposition B. Similarly, the disjunction of any operand B with the Boolean constant "False" yields the proposition B. A truth table (Table 11.2) is employed to define the truth-values of a compound proposition from its constituent propositions. The conjunction of A and B ($A \wedge B$) is true if and only if both A and B are true. The disjunction of A and B ($A \vee B$) is true if either A or B is true.

The "not" operator ($\neg$) is a unary operator such that $\neg A$ is true if A is false and is false if A is true. Complex Boolean expressions may be formed from simple Boolean expressions using the logical connectives (Table 11.3).

---

[4]Finite differences are a numerical method used in solving differential equations.

Table 11.1 Properties
of Boolean algebra

Property	Example
Commutative	$A \wedge B \equiv B \wedge A$
	$A \vee B \equiv B \vee A$
Associative	$A \wedge (B \wedge C) \equiv (A \wedge B) \wedge C$
	$A \vee (B \vee C) \equiv (A \vee B) \vee C$
Identity	$A \wedge \text{True} \equiv A$
	$A \vee \text{False} \equiv A$
Distributive	$A \wedge (B \vee C) \equiv (A \wedge B) \vee (A \wedge C)$
	$A \vee (B \wedge C) \equiv (A \vee B) \wedge (A \vee C)$
De Morgan	$\neg (A \wedge B) \equiv \neg A \vee \neg B$
	$\neg (A \vee B) \equiv \neg A \wedge \neg B$
Idempotent	$A \wedge A \equiv A$
	$A \vee A \equiv A$

Table 11.2 Truth tables for
conjunction and disjunction

A	B	$A \wedge B$	A	B	$A \vee B$
T	T	T	T	T	T
T	F	F	T	F	T
F	T	F	F	T	T
F	F	F	F	F	F

Table 11.3 Truth table for
not operation

A	$\neg A$
T	F
F	T

11.2   Switching Circuits and Boolean Algebra

Claude Shannon showed in his famous Master's thesis that Boolean algebra
provided the perfect mathematical model for switching theory and for the design
of digital circuits. It may be employed to optimize the design of systems of
electromechanical relays, and circuits with relays solve Boolean algebra problems.

Modern electronic computers use millions (billions) of transistors that act as
switches and can change state rapidly. The use of switches to represent binary values
is the foundation of modern computing. A high voltage represents the binary value 1
with low voltage representing the binary value 0. A silicon chip may contain billions
of tiny electronic switches arranged into logical gates. The basic logic gates are
AND, OR and NOT. These gates may be combined in various ways to allow the
computer to perform more complex tasks such as binary arithmetic. Each gate has
binary value inputs and outputs.

The example in Fig. 11.2 is that of an "AND" gate which produces the binary
value 1 as output only if both inputs are 1. Otherwise, the result will be the binary
value 0. Figure 11.3 is an "OR" gate which produces the binary value 1 as output if
any of its inputs is 1. Otherwise, it will produce the binary value 0.

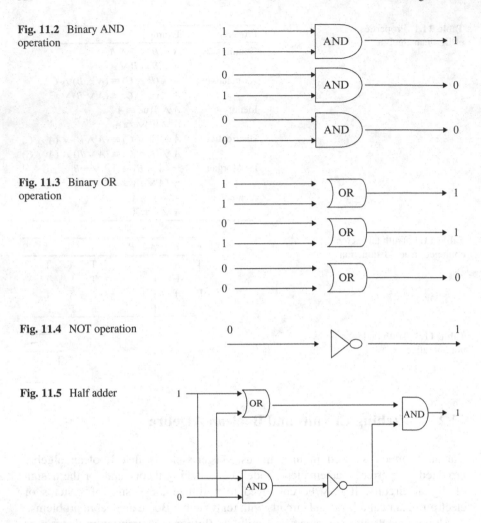

Fig. 11.2 Binary AND operation

Fig. 11.3 Binary OR operation

Fig. 11.4 NOT operation

Fig. 11.5 Half adder

Finally, a NOT gate accepts only a single input which it reverses. That is, if the input is "1", the value "0" is produced and vice versa (Fig. 11.4).

The logic gates may be combined to form more complex circuits. The example in Fig. 11.5 is that of a half adder of $1 + 0$. The inputs to the top OR gate are 1 and 0 which yield the result of 1. The inputs to the bottom AND gate are 1 and 0 which yield the result 0, which is then inverted through the NOT gate to yield binary 1. Finally, the last AND gate receives two 1s as input and the binary value 1 is the result of the addition. The half adder computes the addition of two arbitrary binary digits, but it does not calculate the carry. It may be extended to a full adder that provides a carry for addition.

Chapter 12
Fred Brooks

Fred Brooks is an American computer scientist who has made important contributions to software engineering and project management. He is famous for project managing the IBM/360 project, and his formulation of Brooks' law which states that adding more people to a project that is running late makes the project even later (Fig. 12.1).

He was born in North Carolina in 1931 and studied physics at Duke University in North Carolina. He received his bachelor's degree in 1953 and moved to Harvard University in Massachusetts to do a Ph.D. in computer science. His doctoral advisor was Howard Aiken who had designed the Harvard Mark I. Brooks obtained his Ph.D. in 1956 and he joined IBM in New York later that year.

He worked on the architecture (he actually coined the term *computer architecture*) of various IBM machines including the Stretch and Harvest computers. Brooks and Dura Sweeney patented a program interrupt system for the IBM Stretch computer in 1957, and this system includes many features of today's interrupt systems.

He became project manager for the development of the System/360 family of computers and the IBM OS/360 operating system. The System/360 project involved 5,000 man-years of effort at IBM.

The System/360 project was effective in achieving strict compatibility in the System/360 family of computers, and the project introduced a number of new industry standards including 8-bit bytes.

Brooks recorded his experience as project manager in a famous project management book titled *The Mythical Man Month* [Brk:75]. This book which appeared in 1975 considered the challenge of delivering a major project (of which software is a key constituent) on time, on budget and with the right quality. Brooks described it as *my belated answer to Tom Watson's probing question as to why programming is hard to manage.*

He wrote a famous paper [Brk:86] *No Silver Bullet—Essence and Accidents of Software Engineering* in 1986. This paper argues that "There is no single development in technology or management techniques which by itself promises even one order of magnitude [tenfold] improvement within a decade in productivity,

Fig. 12.1 Fred Brooks
(Photo by Dan Sears)

reliability and simplicity". Brooks states that while there is no silver bullet to achieve this goal, that a series of innovations, could lead to significant improvements, and perhaps greater than tenfold improvements over a 10-year period.

Brooks founded the Department of Computer Science at the University of North Carolina at Chapel Hill in 1964 and chaired it for over 20 years. His recent research is in computer graphics and virtual reality, and he has supervised several Ph.D. students.

He has received many awards for his contributions to the computing field. He received the ACM Turing award in 1999 in recognition of his contribution to software architecture, operating systems and software engineering. He has served on the National Science Board and the Defence Science Board.

12.1   Mythical Man-Month

The Mythical Man-Month is an influential software project management book, and the essays are based on Brooks experience as project manager of the System/360 project. It is his analysis of the management and technical lessons to be learned from this large computer programming project.

This collection of essays have the underlying theme that *large programming projects encounter management problems on a scale that is different to smaller projects due to the division of labour.*

Brooks argues that existing estimation techniques are poor and are based on the false optimistic assumption that everything will go well.[1] Further schedule progress is poorly monitored, and the response to schedule slippage (when it is recognized)

---

[1]Programmers tend to be optimistic with the belief that all will go well: i.e. that each task will take only as long as it ought to take. Inconsistencies and incompleteness of ideas often become apparent only during the implementation.

is often to add more manpower to the project,[2] which makes matters much worse. Brooks fundamentally questions the very unit of effort (i.e. the *man-month*) used in estimation and scheduling. He argues that while cost varies as the product of the number of staff and the number of months, that progress does not. He states that *The man-month as a unit for measuring the size of a job is a dangerous and deceptive myth*, and therefore, it is dangerous and deceptive to view men and months as interchangeable.

He argues that men and months are interchangeable only when a task can be partitioned among several workers with no communication between them. Complex programming projects cannot be perfectly partitioned into discrete tasks to be worked on without communication between the programmers. There is a complex relationship between the tasks and the programmers performing them. A project with n people requires a pairwise intercommunication effort of $n(n - 1)/2$. Clearly, as n increases there is a significant increase in the pairwise intercommunication effort. Some tasks have sequential constraints and cannot be partitioned, and so the only way to improve the schedule is by increased effort.

Brooks' law states *Adding manpower to a late software project makes it later*. Many project managers agree with this statement on schedule slippage (*from bitter experience*). The problem is that project staff will need to devote time to training new team members to become familiar with the technology and the project. This leads to a loss of team productivity, and it takes time for the new team members to become productive and effective. There are also increased communication overheads due to the increase in size of the project team. Communication may take place informally, or involve project meetings, or consist of a collection of documents such as the project notebook.

Brooks highlighted the importance of schedule monitoring with another famous quote: *How does a project get to be a year behind schedule? One day at a time*. This shows the importance of monitoring the project schedule on a daily and weekly basis and taking early action to deal with incremental slippage. It is important to meet the small individual milestone to ensure that the overall schedule remains on track. There are often more effective solutions to schedule slippage other than adding extra staff to the project.

The book includes several essays on architecture and system design, and he argues that architecture must be separated from implementation. The architecture states what the system is to do, whereas implementation states how it is to be done. For example, the System/360 has a single architecture that is implemented quite differently in each of the nine models of the family. The architects need sufficient time to determine the architecture and to document the specification in a manual. Brooks argues that when designing a new system, it will be necessary to design a throw away system that acts as a pilot that will yield valuable feedback to be used in the redesign of the system.

---

[2]Brooks describes the effect of this approach as being similar to dousing a fire with gasoline.

The programmers then implement the architecture and build the system. Brooks discusses the second-system effect, which argues that the second system is the most dangerous in that the architect will tend to include many or all of the ideas that he did not add to the first system due to time constraints.

Brooks notes that *change is inevitable and that it is essential that a system be designed for change*, and that sound configuration management practices are employed. At a certain stage, no more changes should be allowed to the system and the code should be frozen, with change requests deferred to the next version of the system.

12.2   No Silver Bullet

Brooks' famous paper *No Silver Bullet—Essence and Accidents of Software Engineering* was published in 1986 and is a classic in software engineering. Brooks describes the quest for a silver bullet that can magically ensure that a project does not become a monster (or werewolf) with missed schedules, massively exceeded budgets and flawed products. However, he argues that there is no silver bullet in either technology or management technique that would lead to the desired improvements in productivity, reliability and simplicity. Brooks explains why this is the case by considering several silver bullets proposed and the nature of software projects.

He argues that it is unlikely that there will be any inventions that will do for software productivity, reliability and simplicity what electronics, transistors and large-scale integration did for computer hardware with a doubling of hardware productivity roughly every 2 years. Further, it is not that the gains in the software field are slow: rather it is that there have been massive gains in productivity in computer hardware, and that no other technology in civilization has witnessed such productivity and cost improvements.

Brooks then considers the progress that may be expected with software projects by considering difficulties that are inherent in software production (i.e. *essence*), as well as difficulties that are present but are not inherent (i.e. *accidents*).[3]

He identified a number of inherent difficulties with software entities. These include *complexity*, as software systems are *an inherently complex intellectual undertaking*, and this leads to both technical and management challenges. The complexity increases the communication difficulty between team members as well as making it easy to include side effects when introducing new functionality.

---

[3]The classification of an entity into its essence and accidents is due to Aristotle. His metaphysics introduces the doctrine of the categories, where substance (or essence) is the primary form of being and the characteristics of a substance (i.e. accidents) are secondary attributes such as quality, quantity and relation.

Software systems face constant pressure for change from users who wish for extensions to the current functionality to address their specific needs and from changes to the systems that the software is embedded within.

Brooks argues that success to date with difficulties with software projects has dealt with the accidental difficulties rather the essential difficulties. He gives the example of the introduction of high-level programming languages which led to improvements in productivity, reliability and simplicity. However, he argues that this has dealt with accidental complexity rather than essential complexity. Similarly, the introduction of unified programming environments has improved programmer productivity by addressing accidental difficulties rather than essential.

He considered emerging developments that are proposed as potential silver bullets and considered whether they offer a revolutionary or incremental advance. He considered the potential of emerging programming languages such as Ada, object-oriented programming, artificial intelligence, expert systems, automatic programming, graphical programming, program verification, environment and tools and workstations. He argues that while no technology will give the results obtained in the hardware field, there is a lot of good work taking place and lots of steady incremental improvements.

Brooks' conclusion is that while there is no silver bullet to achieve the goal of productivity improvements as in the hardware field, a series of innovations could lead to significant improvements in the software field.

Chapter 13
Vannevar Bush

Vannevar Bush was an American scientist who developed the differential analyzer at MIT. He played a leading role in shaping American policy on scientific research and in developing close links between academia and the military. He was the author of a famous article "As we may think" in 1945, which outlined a vision of the Internet and the World Wide Web (Fig. 13.1).

He was born in Massachusetts in 1890. He obtained a Bachelor of Science and a Master's of Science degree from Tufts College (a private college in Massachusetts) in 1913. He earned his Ph.D. in electrical engineering from the Massachusetts Institute of Technology (MIT) in 1917. He joined the staff at the Department of Electrical Engineering at MIT in 1919 and co-founded the American Appliance Company (now known as Raytheon) in 1922.

Bush and his graduate students commenced the construction of the differential analyzer in 1927. This analog computer could solve first- and second-order differential equations and was applicable to solving many practical problems in physics. The machine had both electrical and mechanical components, and it was funded by the Rockefeller Foundation.

Bush supervised Claude Shannon at MIT, and Shannon's initial work was to improve the differential analyzer. Shannon showed how Boolean algebra could be applied to digital circuit design, and that Boole's symbolic logic was the perfect mathematical model for switching theory and for the subsequent design of digital circuits and computers.

Bush played an important role in shaping American policy on scientific research and in developing close links between the military and universities. He published a famous article in the 1940s, which described a device called the memex. This device was a proto-hypertext computer system, and it was to inspire others in the development of hypertext systems.

He received several awards for his contributions including the Louis E. Levy Medal from the Franklin Institute in 1928 for the development of the differential analyzer. He received the AIEE Edison's Medal in 1943 for his contributions to

Fig. 13.1 Vannevar Bush

Fig. 13.2 Vannevar Bush
with the differential analyzer

advancing electrical engineering and for his service in guiding the US war research
program. He received awards from Presidents Truman, Johnson and Nixon. He died
in Massachusetts in 1974.

13.1   The Differential Analyzer

The differential analyzer was an analog mechanical device that was regarded as one
of the most accurate calculating devices for its time. It was developed by Vannevar
Bush and others at the Massachusetts Institute of Technology (Fig. 13.2).

Its development commenced in 1928 and it was completed in 1931. It was
designed to handle differential equations and solved them by integration, and it used
one or more wheel and disc integrators interconnected by shafts in various ways to
solve the particular problem.

13.2 Bush's Influence on Scientific Research

Bush became president of the Carnegie Institute in 1938, and this high-profile position allowed him to influence research policy in the United States and to advise the government on scientific research. He was appointed to the National Advisory Committee for Aeronautics (NACA) in the same year and became chairman in 1939. NACA was the predecessor of NASA.

He created the National Defense Research Committee (NDRC) in 1940, and this agency helped to improve coordination and communication of scientific research among the military.

Bush became director of the Office of Scientific Research and Development (OSRD) in 1941, and *he developed a win-win relationship between the US military and American universities*. He arranged large research funding for the universities to carry out applied research to assist the US military. This allowed the military to benefit from the early exploitation of research results, and it also led to better facilities and laboratories for the universities.

This led to close links and cooperation between American universities such as Harvard and Berkeley, and these close links later facilitated the development of ARPANET by DARPA.

He played an important role in persuading the United States government to develop the atomic bomb, and he was part of the committee that advised Truman on the use of nuclear weapons. The OSRD was abolished in 1947 and the National Science Foundation was founded in 1950.

13.3 The Memex

The vision of the Internet and World Wide Web goes back to an article by Vannevar Bush in the 1940s. Bush outlined his vision of an information management system called the *memex* (memory extender) in a famous essay "As we may think" [Bus:45].

He envisaged the memex as a device electronically linked to a library and able to display books and films. It describes a proto-hypertext computer system and influenced the later development of hypertext systems.

A memex is a device in which an individual stores all his books, records, and communications, and which is mechanized so that it may be consulted with speed and flexibility. It is an enlarged intimate supplement to his memory.

It consists of a desk, and while it can presumably be operated from a distance, it is primarily the piece of furniture at which he works. On the top are slanting translucent screens, on which material can be projected for convenient reading. There is a keyboard, and sets of buttons and levers. Otherwise it looks like an ordinary desk.

Bush predicted that:

Wholly new forms of encyclopedias will appear, ready made with a mesh of associative trails running through them, ready to be dropped into the memex and there amplified.

Ted Nelson and Douglas Engelbart independently formulate ideas that would become hypertext. Tim Berners-Lee would later use hypertext as part of the development of the World Wide Web.

Chapter 14
Vint Cerf

Vint Cerf is an American computer scientist who is regarded (with Robert Kahn) as one of the fathers of the Internet. He is currently vice president and Internet Evangelist with Google (Fig. 14.1).

He was born in Connecticut in 1943 and studied mathematics at Stanford University. He obtained his bachelor's degree 1965 and later pursued graduate studies at UCLA where he obtained an MS degree in 1970 and a Ph.D. in 1972. He worked with the data packet network group in UCLA, and this group connected the first two nodes in ARPANET (the predecessor of the Internet). He met Robert Kahn during this period, and Kahn was working on ARPANET hardware at UCLA. Cerf was an assistant professor at Stanford University from 1972 to 1976, where he carried out research on packet network interconnection protocols. He codesigned the TCP/IP set of protocols with Kahn during this period, and the transmission control protocol (TCP) and the Internet protocol (IP) are the fundamental protocols at the heart of the Internet.

He was a program manager and principal scientist at DARPA (US Defense Advanced Research Projects Agency) from 1976 to 1982, where he was responsible for the packet technology and network security research programs. He was vice president of engineering at MCI from 1982 to 1986, where he was responsible for the design and implementation of MCI Mail. This was the first commercial email service to be connected to the Internet.

He is currently vice president and chief Internet Evangelist with Google (since 2005) and is responsible for identifying new enabling technologies and applications for the Internet and other platforms for the company.

Cerf and Kahn have received numerous awards for their contributions to the development of the Internet. These include the US National Medal of Technology, which they received from President Clinton in 1997. They received the ACM Turing Award in 2004 for their work on Internet protocols. They received the IEEE Alexander Graham Bell Medal in 1997. Cerf has received several honorary doctorates from universities around the world.

G. O'Regan, *Giants of Computing: A Compendium of Select, Pivotal Pioneers*,
DOI 10.1007/978-1-4471-5340-5_14, © Springer-Verlag London 2013

Fig. 14.1 Vint Cerf

14.1 TCP/IP

There were approximately 10,000 computers in the world in the 1960s. These were expensive machines with limited processing power, and communication between computers was virtually nonexistent. Several computer scientists had dreams of a worldwide network of computers, with every computer around the globe connected to all others. This would allow everyone in the world to be connected and to access programs and data from anywhere.

The US Department of Defense founded the Advanced Research Projects Agency (ARPA) in the late 1950s, as a body to manage the development of new and advanced technologies for the US military. The concept of packet switching[1] was invented in the early 1960s, and several organizations commenced work on its implementation.

The first wide area network connection was created in 1965 and involved the connection of a computer at MIT to a computer in Santa Monica (via a dedicated telephone line). This demonstrated the feasibility of a telephone line for data transfer, and ARPA recognized the need to build a network of computers. This led to the ARPANET project in 1966 which aimed to implement a packet switched network with a network speed of 56 kbps.

One early public demonstration of ARPANET was in 1972, and included a demo of Weizenbaum's famous ELIZA program.[2] This well-known AI program allowed a user to conduct a typed conversation with an artificially intelligent machine (psychiatrist) at MIT.

ARPA was renamed to DARPA in 1972, and it commenced a project to connect several computers in a number of diverse geographical locations using a radio-based network and a project to establish a satellite connection between a site

---

[1] Packet switching is a message communication system between computers. Long messages are split into packets which are then sent separately so as to minimize the risk of congestion.

[2] This program is discussed in a later chapter.

Table 14.1 TCP layers

Layer	Description
Network interface layer	This layer is responsible for formatting packets and placing them on to the underlying network
Internet layer	This layer is responsible for network addressing. It includes the Internet protocol and the address resolution protocol
Transport layer	This layer is concerned with data transport and is implemented by TCP and the user datagram protocol (UDP)
Application layer	This layer is responsible for liaising between user applications and the transport layer. It includes the file transfer protocol (FTP), telnet, domain naming system (DNS) and simple mail transfer program (SMTP)

in Norway and in the UK. This led to a need for the interconnection of the ARPANET with other networks. The key problems were to investigate ways of achieving convergence between the ARPANET, radio-based networks and the satellite networks, as these all had different interfaces, packet sizes and transmission rates. Therefore, there was a need for a network-to-network connection protocol.

The concept of the transmission control protocol (TCP) was developed at DARPA by Vint Cerf and Bob Kahn in 1974 [KaC:74]. TCP is a set of network standards that specify the details of how computers communicate, as well as the standards for interconnecting networks. It allows the internetworking of very different networks, which then act as one network.

It was designed to be flexible and provides a transmission standard that deals with physical differences in host computers, routers and networks. It is designed to transfer data over networks which support different packet sizes and which may sometimes lose packets.

The new protocol standards were known as the *transport control protocol* (TCP) and the *Internet protocol* (IP). TCP details how information is broken into packets and reassembled on delivery, whereas IP is focused on sending the packet across the network. These standards allow users to send electronic mail or to transfer files electronically, without needing to concern themselves with the physical differences in the networks. TCP/IP consists of four layers (Table 14.1):

The Internet protocol (IP) is a connectionless protocol that is responsible for addressing and routing packets. It breaks large packets down into smaller packets when they are travelling through a network that supports smaller packets. A connectionless protocol means that a session is not established before data is exchanged, and packet delivery with IP is not guaranteed as packets may be lost or delivered out of sequence. An acknowledgement is not sent when data is received, and the sender or receiver is not informed when a packet is lost or delivered out of sequence.

A packet is forwarded by the router only if the router knows a route to the destination. Otherwise, it is dropped. Packets are dropped if their checksum is invalid or if their time to live is zero. The acknowledgement of packets is the responsibility of the TCP protocol. The ARPANET employed the TCP/IP protocols as a standard from 1983.

Chapter 15
Alonzo Church

Alonzo Church was an American mathematician and logician who made important contributions to mathematical logic and to theoretical computer science. He developed the lambda calculus in the 1930s as a tool to study computability,[1] and he showed that anything that is computable is computable by the lambda calculus. He proved that the first-order logic is *undecidable* (i.e. there is no algorithm to determine whether an arbitrary mathematical proposition is true or false). He founded the Journal of Symbolic Logic in 1936 (Fig. 15.1).

He was born in Washington, D.C., in 1903 and attended Princeton University. He obtained a bachelor's degree in mathematics in 1924 and earned his Ph.D. in mathematics from the university in 1927. He received a 2-year National Research Fellowship in 1927 and spent a year at Harvard University and a year at the University of Göttingen and the University of Amsterdam. He taught at Princeton from 1929 until his retirement in 1967. He then taught at the University of California from 1967 until his second retirement in 1990. He died in 1995.

15.1 Lambda Calculus

Lambda calculus (λ-calculus) was developed by Alonzo Church in the 1930s to study computability. It is a formal system that may be used to study function definition, function application, parameter passing and recursion. It may be employed to define what a computable function is, and any computable function may be expressed and evaluated using the calculus. Church used lambda calculus in 1936 to give a negative answer to Hilbert's *Entscheidungsproblem*.

---

[1] The Church-Turing thesis states that anytime that is computable is computable by lambda calculus or equivalently by a Turing machine.

G. O'Regan, *Giants of Computing: A Compendium of Select, Pivotal Pioneers*,
DOI 10.1007/978-1-4471-5340-5_15, © Springer-Verlag London 2013

Fig. 15.1 Alonzo Church

It is equivalent to the Turing machine formalism developed by Alan Turing.
However, the emphasis in the lambda calculus is on transformation rules, whereas
Turing machines are concerned with computing on primitive mathematical ma-
chines. Lambda calculus consists of a small set of rules:

- Alpha-conversion rule (α-conversion)[2]
- Beta-reduction rule (β-reduction)[3]
- Eta-conversion (η-conversion)[4]

Every expression in the λ-calculus stands for a function with a single argument.
The argument of the function is itself a function with a single argument, and so on.
The definition of a function is anonymous in the calculus. For example, the normal
definition of a function that adds one to its argument is $f(x) = x + 1$. However, in
λ-calculus this successor function is defined as

$$\lambda x \cdot x + 1$$

The name of the formal argument x is irrelevant, and an equivalent definition
of the function is $\lambda z \cdot z + 1$. The evaluation of a function f with respect to an
argument (e.g. 3) is usually expressed by $f(3)$. In λ-calculus this would be written as
$(\lambda x \cdot x + 1)\ 3$, and this evaluates to $3 + 1 = 4$. Function application is left associative: i.e. $f x y = (f x)\ y$. A function of two variables is expressed in lambda calculus
as a function of one argument which returns a function of one argument. This is
known as *currying*, and it was developed by the Russian logician Schönfinkel and
made popular by the American logician Haskell Curry. For example, the function
$f(x, y) = x + y$ is written as $\lambda x \cdot \lambda y \cdot x + y$. This is often abbreviated to $\lambda x y \cdot x + y$.

---

[2]This essentially expresses that the names of bound variables is unimportant.

[3]This essentially expresses the idea of function application.

[4]This expresses the idea that two functions are equal if and only if they give the same results for
all arguments.

λ-calculus is a simple mathematical system, and its syntax is defined as follows:

$$<exp>::=<identifier> \qquad |$$
$$\lambda<identifier>.<exp> \qquad |-abstraction$$
$$<exp><exp> \qquad |-application$$
$$(<exp>)$$

λ-Calculus's four lines of syntax plus *conversion* rules are sufficient to define Booleans, integers, data structures and computations on them. It inspired Lisp and modern functional programming languages.

15.2 Decidability

Formalism was proposed by Hilbert as a foundation for mathematics in the early twentieth century. A *formal system* consists of a formal language, a set of axioms and rules of inference. Hilbert's program was concerned with the formalization of mathematics (i.e. the axiomatization of mathematics) together with a proof that the axiomatization is consistent. The specific objectives were to:

- Develop a formal system where the truth or falsity of any mathematical statement may be determined.
- Provide a proof that the system is consistent (i.e. that no contradictions may be derived).

A *proof* in a formal system consists of a sequence of formulae, where each formula is either an axiom or derived from one or more preceding formulae in the sequence by an application of one of the rules of inference. Hilbert believed that every mathematical problem could be solved, and he therefore expected that the formal system of mathematics would be *complete* (i.e. all truths could be proved within the system) and *decidable*: i.e. that the truth or falsity of any mathematical proposition could be determined by an algorithm.

He believed that the truth or falsity of any mathematical proposition could be determined in a finite number of steps, and that the formalism would allow a mechanical procedure (or *algorithm*) to determine whether a particular statement was true or false. The problem of the decidability of mathematics is known as the decision problem (*Entscheidungsproblem*).

Definition 15.1 (Decidability) Mathematics is decidable if the truth or falsity of any mathematical proposition may be determined by an algorithm.

Church and Turing independently showed this to be impossible in 1936. Turing showed that decidability was related to the halting problem for Turing machines, and that if first-order logic were decidable, then the halting problem for Turing machines could be solved. However, he had already proved that there was no general algorithm to determine whether a given Turing machine halts or not. Therefore, first-order logic is undecidable.

The question as to whether a given Turing machine halts or not can be formulated as a first-order statement. If a general decision procedure exists for first-order logic, then the statement of whether a given Turing machine halts or not is within the scope of the decision algorithm. However, Turing had already proved that it is not possible algorithmically to decide whether or not an arbitrary Turing machine will halt or not. Therefore, since there is no general algorithm that can decide whether any given Turing machine halts, there is no general decision procedure for first-order logic. The only way to determine whether a statement is true or false is to try to solve it. However, if one tries but does not succeed this does not prove that an answer does not exist.

There are first-order theories that are decidable. However, first-order logic that includes Peano's axioms of arithmetic (or any formal system that includes addition and multiplication) cannot be decided by an algorithm. *Propositional logic is decidable* as there is a procedure (e.g. using a truth table) to determine if an arbitrary formula is valid in the calculus.

15.3   Computability

Church developed the lambda calculus in the mid-1930s, as part of his work into the foundations of mathematics. Turing published a key paper on computability in 1936, which introduced the theoretical machine known as the Turing machine. This machine is computationally equivalent to the lambda calculus and is capable of performing any conceivable mathematical problem that has an algorithm. Church defined the concept of an algorithm formally in 1936 and defined computability in terms of the lambda calculus. Turing defined computability in terms of the theoretical Turing machine.

Definition 15.2 (Algorithm) An algorithm (or effective procedure) is a finite set of unambiguous instructions to perform a specific task.

A function is *computable* if there is an effective procedure or algorithm to compute f for each value of its domain. The algorithm is finite in length and sufficiently detailed so that a person can execute the instructions in the algorithm. The execution of the algorithm will halt in a finite number of steps to produce the value of $f(x)$ for a given x in the domain of f. However, if x is not in the domain of f, then the algorithm may produce an answer saying so, or it may get stuck, or it may run forever and never halt.

The *Church-Turing thesis* states that *any computable function may be computed by a Turing machine*. There is overwhelming evidence in support of this thesis, including the fact that alternative formalizations of computability in terms of lambda calculus, recursive function theory and Post systems have all been shown to be equivalent to Turing machines.

Chapter 16
Noam Chomsky

Noam Chomsky is an American linguist and philosopher and is considered the father of linguistics. He made important contributions to linguistics and to the theory of grammars, and his work has had a major influence on language design and the theory of programming languages.

Chomsky is widely known today as a critic of US foreign policy and as a political commentator. For example, he has argued that the United States *has double standards on foreign policy* in that, on the one hand, it argues for freedom and democracy and, on the other hand, it has maintained close ties with repressive organizations and states, which have led to human rights violations (e.g. links with Chile under Pinochet). Further, he has argued that US intervention in foreign states (e.g. in Nicaragua) is a form of terrorism (Fig. 16.1).

He was born in Philadelphia in 1928 and studied philosophy and linguistics at the University of Pennsylvania. He obtained a bachelor's degree in 1949 and a master's degree in 1951. He earned his Ph.D. from the university in 1955. He became a member of staff at Massachusetts Institute of Technology (MIT) in 1955 and has taught there for over 50 years.

He published an influential book, *Syntactic Structures*, in 1957. This book introduced *transformational generative grammars*, which is a particular approach to the study of syntax. It aims to give a set of rules that will correctly predict which combination of words will form grammatically correct sentences. Chomsky has argued that many of the properties of a grammar arise from an innate *universal grammar* in humans, and there is an evidence for this universal grammar in the language acquisition of children and especially in the speed by which children acquire their native language.

Chomsky argues that a human child has an innate ability (*language acquisition device*) to produce and understand language. He argues that one of the goals of linguistics is to understand this language acquisition device and to determine the constraints it places on possible human languages. The universal features that result from these constraints are termed the *universal grammar* (UG). Chomsky

G. O'Regan, *Giants of Computing: A Compendium of Select, Pivotal Pioneers*,
DOI 10.1007/978-1-4471-5340-5_16, © Springer-Verlag London 2013

Fig. 16.1 Noam Chomsky
(Courtesy of Duncan
Rawlinson)

argues that the grammatical principles underlying grammars are innate and fixed, and differences among the world's languages may be accounted for in terms of parameter settings in the brain.

The formal grammar of a language consists of a finite set of rules and terms and accounts for the ability of a person to create and understand an infinite number of utterances. Panini's grammar of Sanskrit (*Ashtadhyayi*) developed in the first century B.C. is one of the earliest known generative grammars.

Chomsky is the author of over 100 books and has received numerous awards for his contributions. This has included honorary doctorates from many universities around the world.

16.1   Chomsky Hierarchy

The Chomsky hierarchy is important in computer science and has many applications in language design and compiler construction. The theory of the syntax of programming languages is well established, and programming languages have well-defined grammars. A syntactically correct program is generated from the grammar of the programming language.

Chomsky classified grammars into a number of classes with increasing expressive power. The hierarchy consists of four levels including regular grammars, context-free grammars, context-sensitive grammars and unrestricted grammars. Each successive class can generate a broader set of formal languages than the previous. The grammars are distinguished by their production rules, which determine the type of language that is generated (Table 16.1).

Regular grammars are used to generate the words that may appear in a programming language. This includes the identifiers (e.g. names for variables, functions and procedures), special symbols (e.g. addition, multiplication) and the reserved words of the language.

Table 16.1 Chomsky hierarchy of grammars

Grammar type	Description
Type 0 grammar	Type 0 grammars include all formal grammars. They have production rules of the form $\alpha \rightarrow \beta$ where α and β are strings of terminals and nonterminals. They generate all languages that can be recognized by a Turing machine
Type 1 grammar (context sensitive)	These grammars generate the context-sensitive languages. They have production rules of the form $\alpha A\beta \rightarrow \alpha\gamma\beta$ where A is a nonterminal and α, β and γ are strings of terminals and nonterminals. A linear bounded automaton recognizes these languages[a]
Type 2 grammar (context free)	These grammars generate the context-free languages. These are defined by rules of the form $A \rightarrow \gamma$ where A is a nonterminal and γ is a string of terminals and nonterminals. These languages are recognized by a pushdown automaton[b] and are used to define the syntax of most programming languages
Type 3 grammar (regular grammars)	These grammars generate the regular languages (or regular expressions). These are defined by rules of the form $A \rightarrow a$ or $A \rightarrow aB$, where A and B are nonterminals and a is a single terminal. A finite-state automaton recognizes these languages, and regular expressions are used to define the lexical structure of programming languages

[a] A linear bounded automaton is a restricted form of a nondeterministic Turing machine in which a limited finite portion of the tape (a function of the length of the input) may be accessed
[b] A pushdown automaton is a finite automaton that can make use of a stack containing data

A rewriting system for context-free grammars is a finite relation between N and $(A \cup N)^*$, i.e. a subset of $N \times (A \cup N)^*$. A production rule $<N> \rightarrow w$ is an element of this relation and is an ordered pair $(<N>, w)$ where w is a word consisting of zero or more terminal and nonterminal letters. This production rule means that $<N>$ may be replaced by w.

16.2 Computational Linguistics

Linguistics is the theoretical and applied study of language, and human language is highly complex. It includes the study of phonology, morphology, syntax, semantics and pragmatics. Syntax is concerned with the study of the rules of grammar, and the application of the rules of the grammar yields the syntactically valid sentences and phrases.

Morphology is concerned with the formation and alteration of words, and phonetics is concerned with the study of sounds and how sounds are produced and perceived as speech (or nonspeech).

Computational linguistics is a multidisciplinary study of the design and analysis of natural language processing systems. It includes linguists, computer scientists, mathematicians, cognitive psychologists and artificial intelligence experts.

Early work on computational linguistics commenced with machine translation work in the United States in the 1950s. The objective was to develop an automated mechanism by which Russian language texts could be translated directly into English without human intervention. It was naively believed that it was only a matter of time before automated machine translation would be done.

However, the initial results were not very successful, and it was realized that the automated processing of human languages was considerably more complex. This led to the birth of a new field called computational linguistics, and the objective of this field is to investigate and develop algorithms and software for processing natural languages. It is a subfield of artificial intelligence and deals with the comprehension and production of natural languages.

The task of translating one language into another requires an understanding of the grammar of both languages. It requires an understanding of the syntax, the morphology, semantics and pragmatics of the languages. For artificial intelligence to become a reality, it will need to make major breakthroughs in computational linguistics.

Chapter 17
Edgar Codd

Codd was a British mathematician, computer scientist and IBM researcher who developed the *relational database model* in 1970. The relational model is the standard way that information is organized and retrieved from computers, and relational databases are at the heart of systems ranging from hospitals' patient records to airline flight and schedule information (Fig. 17.1).

He was born in Dorset, England, in 1923 and studied mathematics and chemistry at Oxford University. He served as a pilot in the Royal Air Force (RAF) during the Second World War.

He moved to the United States in 1948 to work with IBM as a mathematical programmer and researcher. He initially worked on the SSEC (Selective Sequence Electronic Computer) project in New York and then on the IBM 701 and 702 computers. He spent 3–4 years in Canada in the 1950s working at another company but returned to IBM in 1957 to work on the IBM 7030 STRETCH computer (IBM's first transistorized computer). He was the creator of STEM (statistical database expert manager).

He took leave from IBM from 1961 to 1965 to pursue his Ph.D. studies at the University of Michigan. His thesis was on self-reproducing computers consisting of a large number of identical cells, each of which could interact in a uniform manner with each of its four identical neighbours. This involved building on von Neumann's work in cellular automata, and his Ph.D. was awarded in 1965. He returned to IBM and moved to the IBM laboratory in San Jose, California, in the late 1960s.

He published an internal IBM paper on a relational model in 1969. IBM was promoting the IMS hierarchical database at that time, and it showed little interest or enthusiasm for Codd's model. It made business sense for IBM to preserve revenue for the IMS/DB model rather than embarking on a new technology. However, IBM agreed to implement Codd's ideas on the relational model in the *System-R research project* in the 1970s, and this project demonstrated the power of the model, as well as demonstrating good transaction processing performance. The project introduced a data query language that was initially called SEQUEL (later renamed to SQL), and this language was designed to retrieve and manipulate data in the IBM database.

G. O'Regan, *Giants of Computing: A Compendium of Select, Pivotal Pioneers*,
DOI 10.1007/978-1-4471-5340-5_17, © Springer-Verlag London 2013

Fig. 17.1 Edgar Codd

Codd continued to develop and extend his relational model, and several theorems are named after him. In later years, he proposed a three-valued logic to deal with missing or undefined information and even proposed a four-valued logic in the 1990s. These proposals were never implemented and were controversial at the time.

The relational model became popular from the early 1980s. Codd retired from IBM in 1984 and set up a consulting company to provide education and consultancy services to users of database management systems. He died in Florida in 2003.

He received several awards for his contributions to the computing field. He became an IBM fellow in 1976 and received the ACM Turing Award in 1981 in recognition of his development of the relational model for databases.

17.1   The Relational Model

The concept of a relational database was first described in a paper "A relational model of data for large shared data banks" by Codd [Cod:70]. A relational database is a database that conforms to the relational model, and it may be defined as a set of relations (or tables). The existing database models at the time were the *hierarchical model* and the *network model*.

A binary relation $R(A,B)$ where A and B are sets is a subset of the Cartesian product $(A \times B)$ of A and B. The domain of the relation is A, and the codomain of the relation is B. The notation aRb signifies that there is a relation between a and b and that $(a,b) \in R$. An n-ary relation $R (A_1, A_2, \ldots A_n)$ is a subset of the Cartesian product of the n sets, i.e. a subset of $(A_1 \times A_2 \times \cdots \times A_n)$. However, an n-ary relation may also be regarded as a binary relation $R(A,B)$ with $A = A_1 \times A_2 \times \cdots \times A_{n-1}$ and $B = A_n$.

The data in the relational model are represented as a mathematical n-ary relation. In other words, a relation is defined as a set of n-tuples and is usually represented by

P#	PName	Colour	Weight	City
P1	Nut	Red	12	London
P2	Bolt	Green	17	Paris
P3	Screw	Blue	17	Rome
P4	Screw	Red	14	London
P5	Cam	Blue	12	Paris
P6	Cog	Red	19	London

Fig. 17.2 PART relation

Fig. 17.3 Domains vs. attributes

```
DOMAIN  PART_NUMBER          CHARACTER(6)
DOMAIN  PART_NAME            CHARACTER(20)
DOMAIN  COLOUR               CHARACTER(6)
DOMAIN  WEIGHT               NUMERIC(4)
DOMAIN  LOCATION             CHARACTER(15)

RELATION  PART
    (P#                  : DOMAIN  PART_NUMBER
     PNAME               : DOMAIN  PART_NAME
     COLOUR              : DOMAIN  COLOUR
     WEIGHT              : DOMAIN  WEIGHT
     CITY                : DOMAIN  LOCATION)
```

a table. A table is a visual representation of the relation, and the data is organized in rows and columns. The data stored in each column of the table is of the same data type.

The basic relational building block is the domain or data type (often called just *type*). Each row of the table represents one *n*-tuple (one tuple) of the relation, and the number of tuples in the relation is the cardinality of the relation. Consider the PART relation taken from [Dat:81], where this relation consists of a heading and the body. There are five data types representing part numbers, part names, part colours, part weights, and locations in which the parts are stored. The body consists of a set of *n*-tuples. The PART relation is of cardinality six (Fig. 17.2).

Strictly speaking there is no ordering defined among the tuples of a relation, since a relation is a set, and sets are not ordered. However, in practice, relations are often considered to have an ordering.

There is a distinction between a domain and the columns (or attributes) that are drawn from that domain. An *attribute* represents the *use* of a domain within a relation, and the distinction is often emphasized by giving attributes names that are distinct from the underlying domain. The difference between domains and attributes can be seen in the PART relation (Fig. 17.3) from [Dat:81].

A *normalized relation* satisfies the property that at every row and column position in the table, there is exactly one value (i.e. never a set of values). All relations in a relational database are required to satisfy this condition, and an un-normalized relation may be converted into an equivalent normalized form.

It is often the case that within a given relation, there is one attribute with values that are unique within the relation and can thus be used to identify the tuples of the relation. For example, the attribute P# of the PART relation has this property since each PART tuple contains a distinct P# value, which may be used to distinguish that tuple from all other tuples in the relation. P# is termed the *primary key* for the PART relation. A candidate key that is not the primary key is termed the *alternate key*.

An index is a way of providing quicker access to the data in a relational database, as it allows the tuple in a relation to be looked up directly (using the index) rather than checking all of the tuples in the relation.

The consistency of a relational database is enforced by a set of constraints that provide restrictions on the kinds of data that may be stored in the relations. The constraints are declared as part of the logical schema and are enforced by the database management system (DBMS). They are used to implement the business rules into the database.

17.2   Structured Query Language (SQL)

Codd proposed the Alpha language as the database language for his relational model. However, IBM's implementation of his relational model in the System-R project introduced a data query language that was initially called SEQUEL (later renamed to SQL). This language did not adhere to Codd's relational model but became the most popular and widely used database language. It was designed to retrieve and manipulate data in the IBM database, and its operations included *insert, delete, update, query*, schema creation and modification and data access control.

Structured Query Language (SQL) is a computer language that tells the relational database what to retrieve and how to display it. It was designed and developed at IBM by Donald Chamberlin and Raymond Boyce, and it became an ISO standard in 1987.

A relational database management system (RDBMS) is a system that manages data using the relational model, and examples of such systems include RDMS developed at MIT in the 1970s; Ingres developed at the University of California, Berkeley, in the mid-1970s; Oracle developed in the late 1970s; DB2; Informix; and Microsoft SQL Server.

The most common operation in SQL is the query command, which is performed with the SELECT statement. The SELECT statement retrieves data from one or more tables, and the query specifies one or more columns to be included in the result. Consider the example of a query that returns a list of expensive books (defined as books that cost more than 100.00).

```
SELECT*1
    FROM Book
    WHERE Price > 100.00
    ORDER by title;
```

The *Data Manipulation Language* (DML) is the subset of SQL used to add, update and delete data. It includes the INSERT, UPDATE and DELETE commands. The *Data Definition Language* (DDL) manages table and index structure and includes the CREATE, ALTER, RENAME and DROP statements.

There are extensions to standard SQL that add programming language functionality. A stored procedure is an executable code that is associated with the database. It is usually written in an imperative programming language, and it is used to perform common operations on the database.

Oracle is recognized as a world leader in relational database technology, and its products play a key role in business computing. An Oracle database consists of a collection of data managed by an Oracle database management system. Today, Oracle is the main standard for database technology and is used by companies worldwide.

The success of the Oracle database led to competitor products from IBM (DB2), Sybase (later taken over by SAP), Informix (later taken over by IBM) and Microsoft (SQL server which was developed from Sybase software).

---

[1]The asterisk (*) indicates that all columns of the Book table should be included in the result.

Chapter 18
René Descartes

René Descartes was an influential French mathematician and philosopher. He was born in a village in the Loire Valley in France in 1596 and studied law at the University of Poitiers. He never practised as a lawyer and instead served Prince Maurice of Nassau in the Netherlands. He became interested in mathematics and later invented the Cartesian coordinate system that is used in plane geometry and algebra (Fig. 18.1).

He made important contributions to philosophy and attempted to derive a fundamental set of principles that can be known to be true. His approach was to renounce any idea that could be doubted. He rejected the senses since they can deceive and therefore are not a sound source of knowledge. For example, during a dream, the subject perceives stimuli that appear to be real, but these have no existence outside the subject's mind. Therefore, it is inappropriate to rely on one's senses as the foundation of knowledge.

He argued that a powerful *evil demon or mad scientist* could exist who sets out to manipulate and deceive subjects, thereby preventing them from knowing the true nature of reality. The evil demon could bring the subject into existence with an implanted memory. The question is how one can know for certain what is true given the limitations of the senses. The *brain in the vat thought experiment* is a more modern formulation of the idea of an evil demon or mad scientist. A mad scientist could remove a person's brain from their body and place it in a vat and connect its neurons by wires to a supercomputer. The computer provides the disembodied brain with the electrical impulses that the brain would normally receive. The computer could then simulate reality, and the disembodied brain would have conscious experiences and would receive the same impulses as if it were inside a person's skull. There is no way to tell whether the brain is inside the vat or inside a person.

That is, at any moment, an individual could potentially be a brain connected to a sophisticated computer program or inside a person's skull. Therefore, since you cannot be sure that you are not a brain in a vat, then you cannot rule out the possibility that all of your beliefs about the externalworld are false. The perception

G. O'Regan, *Giants of Computing: A Compendium of Select, Pivotal Pioneers*,
DOI 10.1007/978-1-4471-5340-5_18, © Springer-Verlag London 2013

Fig. 18.1 Rene Descartes

Fig. 18.2 Brain in a VAT
thought experiment

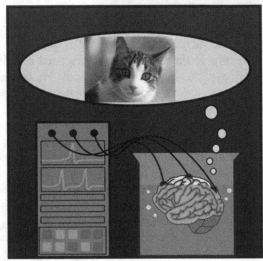

of a *cat* in the case where the brain is in the vat is false and does not correspond to reality. It is impossible to know whether your brain is in a vat or inside your skull. Therefore, it is impossible to know whether your belief is true or false (Fig. 18.2).

Descartes deduced that there is one single principle that must be true. He argued that even if he is being deceived, then clearly he is thinking and must exist. This principle of existence or being is more famously known as *cogito, ergo sum* (*I think, therefore I am*). Descartes argued that this existence can be applied to the present only, as memory may be manipulated and therefore doubted. Further, the only existence that he sure of is that he is a *thinking thing*. He cannot be sure of the existence of his body as his body is perceived by his senses which he has proven to be unreliable. Therefore, his mind or thinking thing is the only thing about him that cannot be doubted. His mind is used to make judgements and to deal with unreliable perceptions received via the senses.

18.1 Cartesian Dualism and AI

Descartes constructed a system of knowledge from his principle of existence using the deductive method. He deduced the existence of a benevolent God using the ontological argument.[1] He argues [Des:99] that we have an innate idea of a supremely perfect being (God), and that God's existence may be inferred immediately from the innate idea of a supremely perfect being.

1. I have an innate idea of a supremely perfect being (i.e. God).
2. Necessarily, existence is perfection.
3. Therefore, God exists.

He then argued that since God is benevolent, he can have some trust in the reality that his senses provide. God has provided him with a thinking mind and does not wish to deceive him. He argued that knowledge of the external world can be obtained by both perception and deduction, and that reason or *rationalism* is the only reliable method of obtaining knowledge. His proof of the existence of God and the external world are controversial.

Descartes was a *dualist* and he makes a clear *mind-body* distinction. He states that there are two substances in the universe: mental substances and bodily substances. The mind-body distinction is very relevant in the AI field, and the *analogy of the mind in AI is an AI program running on a computer*, with knowledge gained by sense perception with sensors and logical deduction.

This thinking thing (*res cogitans* or mind/soul) is distinct from the rest of nature (*res extensa*) and interacts with the world through the senses to gain knowledge. Knowledge is gained by mental operations using the deductive method, where starting from the premises that are known to be true, further truths may be logically deduced. Descartes founded what would become known as the *rationalist school of philosophy* where knowledge is derived solely by human reasoning.

Descartes believed that the bodies of animals are complex living machines without feelings. He dissected (*including vivisection*) many animals for experiments. His experiments led him to believe that the actions and behaviour of non-human animals can be fully accounted for by mechanistic means, and without reference to the operations of the mind. He realized from his experiments that a lot of human behaviour (e.g. physiological functions and blinking) is like that of animals in that it has a mechanistic explanation.

Descartes was of the view that well-designed automata[2] could mimic many parts of human behaviour. He argued that the key differentiators between human and animal behaviour are that humans may adapt to widely varying situations, and that they also have the ability to use language which illustrates the power of thought in

---

[1]The first ontological argument was proposed by St. Anselm of Canterbury in 1078.

[2]An automaton is a self-operating machine or mechanism that behaves and responds in a mechanical way.

humans as distinct from animals. This, he argues, provides evidence for the presence of a soul associated with the human body. In essence, animals are pure machines, whereas humans are machines with minds (or souls).

The significance of Descartes in the field of artificial intelligence is that the Cartesian dualism that humans seem to possess needs to be reflected among artificial machines. Humans seem to have a distinct sense of "I" as distinct from the body, and the "I" seems to represent some core essence of being that is unchanged throughout the person's life. It somehow represents personhood, as distinct from the physical characteristics of a person that are inherited genetically. The challenge for the AI community in the longer term is to construct a machine that (in a sense) possesses Cartesian dualism. That is, the long-term[3] goal of AI is to produce a machine that has awareness of itself as well as its environment.

18.2 Cartesian Geometry

Descartes developed Cartesian geometry in the seventeenth century, and in this system every point in the plane is given an x-coordinate representing its horizontal position, and a y-coordinate representing its vertical position (Fig. 18.3). A point is thus written as an ordered pair (x, y). This system can also be used for three-dimensional geometry where triples (x, y, z) are employed.

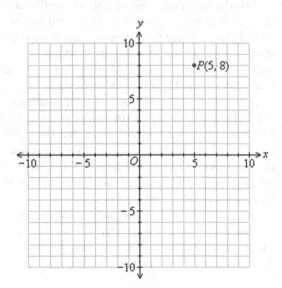

Fig. 18.3 Cartesian Plane

---

[3]This long-term goal may be hundreds of years as there is unlikely to be an early breakthrough in machine intelligence as there are deep philosophical problems to be solved. It took the human species hundreds of thousands of years to evolve to its current levels of intelligence.

Cartesian geometry provides a systematic link between Euclidean geometry and algebra, and it allows geometric shapes such as curves to be described by algebraic equations. For example, the equation of a circle with radius r that is centred at the origin $(0, 0)$ is given by $x^2 + y^2 = r^2$ where (x, y) are the points on the circle.

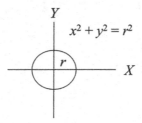

Cartesian coordinates enable the distance between two points to be easily determined. For example, the distance between point $P(x_1, y_1)$ and $Q(x_2, y_2)$ in the plane is given by

$$d = \sqrt{(x_2 - x_1)^2 + (y_2 - y_1)^2}.$$

Cartesian coordinates enable the graph of a function to be easily sketched, and it is an essential tool in mathematics and science. It may be extended to n-dimensional space, where each point in n-dimensional space is represented by $(x_1, x_2, \ldots, x_n)$. There are several other coordinate systems that have been developed including polar coordinates and spherical coordinates.

Chapter 19
Tom DeMarco

DeMarco is an American computer scientist, software engineer and author who has made important contributions to project management and software engineering. He was one of the developers of structured analysis in the 1980s (Fig. 19.1).

He was born in Pennsylvania in 1940 and studied electrical engineering at Cornell University. He obtained a bachelor's degree in 1963 and received a Master's degree from Columbia University in 1965. He obtained a diploma from the University of Paris at the Sorbonne.

He commenced working at Bell Laboratories in 1963 and participated on the ESS-1 project team. This was the first large-scale Electronic Switching System, and it was installed in telephone offices throughout the world. He then worked for La CEGOS Informatique, a French management consulting company, where he was responsible for managing real-time projects and distributed online banking systems in Europe.

He was one of the founders of Atlantic Systems Guild in the 1980s. This consulting company was based in New York and London and specialized in methods and management of software development. He remains a principal consultant with the Atlantic Systems Guild and is also a senior consultant with the Cutter Consortium. He has lectured and consulted in organization development and system development methodologies around the world.

His areas of interests include project management, organization development and culture and change facilitation. He is the author of several books on management, organization design and systems development, as well as books on fiction. His books include well-known titles such as *Structured Analysis and System Development*; *Controlling Software Projects*: *Management, Measurement and Estimates*; *Why Does Software Cost So Much?*; and *Peopleware, Productive Projects and Teams*.

He has received various awards for his contributions to software development methods. He was the recipient of the *Warnier Prize* in 1986 for lifelong contributions to information sciences. He received the *Stevens Award* in 1999 for outstanding contributions to methods for software and system development.

G. O'Regan, *Giants of Computing: A Compendium of Select, Pivotal Pioneers*, DOI 10.1007/978-1-4471-5340-5_19, © Springer-Verlag London 2013

Fig. 19.1 Tom DeMarco

19.1   Structured Analysis and Data Flow Diagrams

Structured analysis is a collection of analysis, design and programming techniques designed to address problems with software development in the late 1960s. There was little guidance at the time on sound techniques to design and develop software, and no agreed techniques for specifying and documenting requirements and design. Structured analysis views the system from the point of view of data flowing through the system and uses data flow diagrams.

DeMarco has stressed the importance of maintaining the specification, since a legacy information system may consist of several million lines of code, written and enhanced over many years by employees who are no longer with the company. He argues that nobody now knows what the system does, as program listings and source code have been lost and the specifications are completely out of date. Further, the system has grown so large that neither the users nor the data processing people have the faintest idea of what the system is supposed to be doing. He argues that this is the fate of all large systems, unless steps are taken to keep the specifications up to date for the lifetime of the system.

He has advocated the approach known as structured analysis [DeM:79], which views a system from the point of view of the data flowing through the system. The function of the system is described by the processes that transform the data flows. It provides a top-down graphic model of the system-to-be, and explains how an analyst proceeds from a physical description of the user's current system to a logical description of the current system, and finally to a logical description of the system-to-be. He stresses the importance of a data dictionary and mini-specs written in structured English. His approach consists of:

- Context diagram
- Data flow diagram
- Process specification
- Data dictionary

Fig. 19.2 DeMarco symbols in DFD

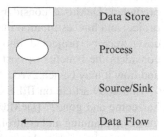

The *context-level* data flow diagram shows the interaction between the system and the external agents that act as *data sources* and *data sinks*. It shows the system boundaries, the external entities that interact with the system and the major information flows between the entities and the system. The system's interactions with the outside world are modelled in terms of data flows across the system boundary. The context diagram shows the entire system as a *single process* without the details of its internal organization.

The context-level diagram is then expanded to produce a *level 0 data flow diagram*. This shows the system's major processes, data flows and data stores at a high level of detail. It shows how the system is divided into processes, each of which deals with one or more of the data flows to or from an external agent, giving the functionality of the system as a whole. The processes are labelled 1.0, 2.0, etc., and these are then decomposed into lower level DFDs.

A *data flow diagram* provides a graphical representation of the flow of data between processes in an information system and is used by the analyst to give an overview of the system (level 0), which may then be expanded into the more detailed levels. The data flow diagram allows the user to visualize how the system will operate and what the system will accomplish.

Data flow diagrams have a hierarchy, and a diagram may consist of several layers each unique to a specific process. Data flow diagrams employ symbols to illustrate data movement and sources and destination of data in a system. The symbols are described in Fig. 19.2.

A process definition is needed to capture the transaction/transformation information. Data dictionaries are needed to describe the data and command flows.

19.2 Management and Organization Development

DeMarco has made important contributions to the management aspects of software development. One of his early books is *Controlling Software Projects: Management, Measurement and Estimates* [DeM:86], and this book opens with the famous quote *You can't control what you can't measure.*

The focus of the book is to develop an effective strategy to measuring software development costs, as this allows the costs of future projects to be more accurately predicted. DeMarco considers what it means to control a software development project and how estimation may be employed effectively. He examines how project costs may be projected into future and what metrics should be collected. He considers the benefits of cost models and examines what software quality means and how it may be achieved.

In a 2009 article on IEEE Software "Software engineering: an idea whose time has come and gone?" [DeM:09], DeMarco seems to challenge some of his earlier views on planning and control. He argues for an emphasis on a return on investment from the software development and in developing software that changes the world or transforms business practices. Further, he adds that while software engineering is a desirable goal, there is always an experimental aspect to software development.

DeMarco and Lister wrote *Peopleware*: *Productive Projects and Teams* [DeM:99] in the late 1980s. This is a popular project management book that argues the challenges of project management are more sociological in nature rather than due to the complexities of the technology. They argue that *the manager's job is not to make people work, but to make it possible for people to work.*

Chapter 20
Edsger Dijkstra

Edsger Dijkstra was a famous Dutch computer scientist who made important contributions to language development, operating systems, graph theory and formal program development (Fig. 20.1).

He was born in Rotterdam in Holland in 1930, and he studied mathematics and physics at the University of Leiden. He obtained a Ph.D. in computer science from the University of Amsterdam in 1959 and commenced his programming career at the Mathematics Centre in Amsterdam in the early 1950s. He developed various efficient graph algorithms to determine the *shortest* or *longest* paths from a vertex u to a vertex v in a graph.

He contributed to the definition of Algol 60 and was part of the team that implemented the first Algol 60 compiler. He became a professor of mathematics at Eindhoven University in the early 1960s. He has made many contributions to computer science, including language development, operating systems and formal program development. Some of his achievements are summarized below (Table 20.1).

The NATO software engineering conference in 1968 highlighted a *software crisis* with projects being delivered late and with poor quality. This led to an interest in rigorous techniques to engineer software.

Dijkstra's *Go to statement considered harmful* article was influential in the trend towards structured programming, and he considered testing to be an inappropriate means of building quality into software: *Testing a program shows that it contains errors never that it is correct* [Dij:72].

He advocated simplicity, precision and a formal approach to program development using the calculus of weakest preconditions. He insisted that programs should be composed correctly using mathematical techniques and not debugged into correctness.

His calculus of weakest preconditions is used to develop reliable programs, and his approach is termed *program derivation*. The program is shown to be correct by construction, and it involves *developing the program and its proof hand in hand*. The proof uses weakest preconditions and the formal definition of the programming language constructs.

G. O'Regan, *Giants of Computing: A Compendium of Select, Pivotal Pioneers*, 91
DOI 10.1007/978-1-4471-5340-5_20, © Springer-Verlag London 2013

Fig. 20.1 Edsger Dijkstra
(Courtesy of Brian Randell)

Table 20.1 Dijkstra's achievements

Area	Description
Go To statement	Dijkstra argued against the use of the *goto* statement in programming
Graph algorithms	He developed efficient graph algorithms to determine the *shortest* (Dijkstra's algorithm) or *longest* paths from a vertex u to vertex v in a graph
Operating systems	He discovered that operating systems can be built as synchronized sequential processes. He introduced key concepts such as *semaphores* and *deadly embrace*
Algol 60	He contributed to the definition of the language and was part of the team that developed the first Algol 60 compiler
Formal program development (guarded commands)	He introduced *guarded commands* and *predicate transformers* as a means of defining the semantics of a programming language. He showed how *weakest preconditions* can be used as a calculus (*wp*-calculus) to develop reliable programs

He viewed programming as a *goal-oriented activity* with the desired result (i.e. the *postcondition R*) playing a more important role in the development of the program than the *precondition Q*. That is, programming is employed to solve a particular problem (or goal), and the problem needs to be clearly stated with precise *pre-* and *post*conditions.

He received the Turing award in 1972 for fundamental contributions to developing programming languages. He moved to the University of Texas in 1984, where he held the Schlumberger Centennial Chair in Computer Science until his retirement in 2000. He died of cancer in 2002. Dijkstra's calculus of weakest preconditions is described in the next section.

20.1 Calculus of Weakest Preconditions

Dijkstra's calculus of weakest preconditions [Dij:76] is applied to the formal development of programs. The weakest precondition $wp(S, R)$ is a predicate that describes a set of states. S represents a command and R represents the postcondition. It is defined as follows:

Definition (Weakest Precondition) The predicate wp(S, R) represents the set of all states such that if execution of S commences in any one of them, then it is guaranteed to terminate in a state satisfying R.

For example, let S be the assignment command $i := i + 5$ and let R be $i \leq 3$, then

$$wp(i := i + 5; \ i \leq 3) = (i \leq -2)$$

The weakest precondition $wp(S,T)$ represents the set of all states such that if execution of S commences in any one of them, then it is guaranteed to terminate.

$$wp \ (i := i + 5; \ T) = T$$

The weakest precondition $wp(S,R)$ is a precondition of S with respect to R and is the weakest such precondition. For any other precondition P of S with respect to R, then $P \Rightarrow wp(S, R)$.

For a fixed command S, then $wp(S, R)$ can be written as a function of one argument $wp_S(R)$, and the function wp_S transforms the predicate R to another predicate $wp_S(R)$; i.e. the function wp_S acts as a *predicate transformer*.

An imperative program may be regarded as a predicate transformer. A predicate P characterizes the set of states in which the predicate P is true, and so an imperative program may be regarded as a binary relation on states, which may be extended to a function F, leading to the Hoare triple $P\{F\}Q$. That is, the program F acts as a predicate transformer.

The predicate P may be regarded as an input assertion and is true before the program F is executed. The predicate Q is the output assertion and is true if the program F terminates, having commenced in a state satisfying P.

20.1.1 Properties of Weakest Preconditions

The weakest precondition $wp(S,R)$ has several well-behaved properties such as (Table 20.2):

Table 20.2 Properties of WP

Law	Description
Excluded miracle $wp\,(S, F) = F$	This describes the set of states such that if execution commences in one of them, then it is guaranteed to terminate in a state satisfying false. However, no state ever satisfies false, and therefore $wp(S, F) = F$. (In other words, it would be a miracle if execution could terminate in no state.)
Distributive $wp\,(S, Q) \wedge wp\,(S, R) =$ $\quad wp\,(S, Q \wedge R)$	This property stipulates that the set of states such that if execution commences in one of them, then it is guaranteed to terminate in a state satisfying $Q \wedge R$ is precisely the set of states such that if execution commences in one of them, then execution terminates with both Q and R satisfied
Monotonicity $Q \Rightarrow R$ then $wp\,(S, Q) \Rightarrow wp\,(S, R)$	This property states that if a postcondition Q is stronger than a postcondition R, then the weakest precondition of S with respect to Q is stronger than the weakest precondition of S with respect to R
Disjunction $wp\,(S, Q) \vee wp\,(S, R) \Rightarrow$ $\quad wp\,(S, Q \vee R)$	This property states that the set of states corresponding to the weakest precondition of S with respect to Q or the set of states corresponding to the weakest precondition of S with respect to R is stronger than the weakest precondition of S with respect to $Q \vee R$. (Equality holds only when the execution of the command is deterministic)

Weakest Preconditions of Commands

The weakest precondition can be used to provide the definition of commands in a programming language. The commands considered here in Table 20.3 are as in [Gri:81].

The **cand** operator was introduced by Dijkstra to deal with undefined values, and the expression a **cand** b is equivalent to **if** a **then** b **else** F. Each $B_i \rightarrow S_i$ is a guarded command (S_i is any command). The guards must be well-defined in the state where execution begins, and at least one of the guards must be true or execution aborts. If at least one guard is true, then one guarded command $B_i \rightarrow S_i$ with true guard B_i is chosen and S_i is executed.

The meaning of the iterate command is that a guard B_i is chosen that is true and the corresponding command S_i is executed. The process is repeated until there are no more true guards. Each choice of a guard and execution of the corresponding statement is an iteration of the loop. On termination of the iteration command, all of the guards are false.

Table 20.3 Properties of WP

Command	Description
Skip command $wp\,(skip, R) = R$	The *skip* command does nothing and is used to explicitly say that nothing should be done. The predicate transformer wp_{skip} is the identity function
Abort command $wp\,(abort, R) = F$	The *abort* command is executed in a state satisfying false (i.e. no state). If program execution reaches a point where *abort* is to be executed, then it is in error. This command should never be executed
Sequential composition $wp\,(S_1; S_2, R) =$ $\quad wp\,(S_1, wp\,(S_2, R))$	The sequential composition command composes two commands S_1 and S_2 by first executing S_1 and then executing S_2. Sequential composition is expressed by $S_1; S_2$. It is associative: $wp\,(S_1; (S_2; S_3), R) = wp\,((S_1; S_2); S_3, R)$
Simple assignment $wp\,(x := e, R) =$ $\quad dom(e) \; \mathbf{cand} \; R_e^x$	The execution of the *assignment* command consists of evaluating the value of the expression e and storing its value in the variable x. It may be executed only in a state where e may be evaluated
Often, the domain predicate $dom(e)$ that describes the set of states that e may be evaluated is omitted and so $wp\,(x := e, R) = R_e^x$	The expression R_e^x denotes the expression obtained by substituting e for all free occurrences of x in R e.g. $(x + y > 2)_v^x = v + y > 2$ In other words, R will be true after execution if and only if the predicate R with the value of x replaced by e is true before execution (since x will contain the value of e after execution)
Assignment to array element $wp\,(b[i] := e, R) = inrange\,(b, i)_a$ $\quad \mathbf{cand}\,dom(e) \; \mathbf{cand} \; R_{(b;i:e)}^b$	The execution of the *assignment to an array element* command consists of evaluating the expression e and storing its value in the array element subscripted by i The $inrange(b,i)$ and $dom(e)$ are often omitted in practice and so $wp\,(b[i] := e, R) = R_{(b;i:e)}^b$
Alternate command $wp\,(IF, R) =$ $\quad dom\,(B_1 \vee B_2 \vee \ldots \vee B_n)$ $\quad \wedge\,(B_1 \vee B_2 \vee \ldots \vee B_n)$ $\quad \wedge\,(B_1 \Rightarrow wp\,(S_1, R))$ $\quad \wedge\,(B_2 \Rightarrow wp\,(S_2, R))$ $\quad \wedge \ldots \wedge (B_n \Rightarrow wp\,(S_n, R))$	The *alternate* command is the familiar **if** statement of programming languages. Its general form is **If** $B_1 \to S_1$ $\quad B_2 \to S_2$ $\quad \ldots$ $\quad B_n \to S_n$ **fi**

(continued)

Table 20.3 (continued)

Command	Description
Iterative command $wp\,(DO, R) =$ $(\exists k : 0 \le k : H_k(R))$ where $H_k(R)$ is defined as $H_k(R) =$ $H_0(R) \lor wp\,(\text{IF}, H_{k-1}(R))$	The *iterate* command is the familiar, while loop statement of programming languages. Its general form is $$\mathbf{do}\,B_1 \rightarrow S_1$$ $$B_2 \rightarrow S_2$$ $$\dots$$ $$B_n \leftarrow S_n$$ $$\mathbf{od}$$

[a] The notation $(b;i:e)$ denotes an array identical to array b except that the array element subscripted by i contains the value e

20.1.2   Formal Program Development with WP

The use of weakest preconditions for formal program development is described in [Gri:81]. Its approach is to develop the program and its proof hand in hand, and the resulting program is then correct by construction.

Programming is viewed as a goal-oriented activity in that the desired result (i.e. the postcondition R) plays a more important role than the precondition Q. Consider the example of a program P to determine the maximum of two integers x and y [Gri:81]. A program P is required that satisfies

$$\{T\}\,P\,\{R : z = \max(x, y)\}$$

The postcondition R is then refined by replacing max with its definition:

$$\{R : z \ge x \land z \ge y \land (z = x \lor z = y)\}$$

The next step is to identify a command that could be executed in order to establish the postcondition R. One possibility is $z := x$ and the conditions under which this assignment establishes R is given by

$$wp\,(\text{"z} := x\text{"}, R) = x \ge x \land x \ge y \land (x = x \lor x = y)$$
$$= x \ge y$$

Another possibility is $z := y$ and the conditions under which this assignment establishes R is given by

$$wp\,(\text{"z} := y\text{"}, R) = y \ge x$$

The desired program is then given by

$$\textbf{if } x \geq y \rightarrow z := x$$
$$y \geq x \rightarrow z := y$$
$$\textbf{fi}$$

The reader is referred to [Gri:81] for more detailed information and examples of formal program development using the calculus of weakest preconditions.

Chapter 21
George Devol

George Devol was a prolific American inventor and is regarded (with Joseph Engelberger) as one of the fathers of robotics. He was awarded the patent for the first industrial robot (*Unimate*), and he played an important role in the foundation of the modern robotics industry (Fig. 21.1).

He was born in Kentucky in 1912 and was interested in electrical and mechanical devices from an early age. He was an avid reader on everything about mechanical devices, and his initial interests were in the practical application of vacuum tubes. He was not strong academically and did not pursue a university education.

He formed his own company, United Cinephone, in 1932, and one of his early inventions was the automatic door. His company licensed this invention to Yale & Towne who manufactured the "Phantom Doorman" photoelectric door. His company also developed a bar code system that was used for sorting packages at the Railway Express Company.

He sold his company at the start of the Second World War and became a manager at the Special Projects Department at Sperry Gyroscope. This department developed radar devices and microwave test equipment.

After the war, Devol was part of the team that developed the first commercial use of microwave oven technology. He applied for a patent on a magnetic recording system for controlling machines and a digital playback device for machines in 1946. He licensed this digital magnetic recording device to Remington and became manager of their magnetics department. The goal was to develop his device for business data applications, but it proved to be too slow for business data.

He applied for a patent on Programmed Article Transfer in 1954, and the goal of this invention was to perform repeated tasks with greater precision and productivity than a human worker. The patent was concerned with automatic operation of machinery, including handling and an automated control apparatus. It introduced the concept of *universal automation*, and the term *Unimate* was coined. This was the first patent for a digitally operated programmable robot arm, and it led to the foundation of the robotics industry (Fig. 21.2).

Engelberger and Devol founded the first robotics company, *Unimation Inc.*, in 1956. This was the largest robotics company in the world for many years.

G. O'Regan, *Giants of Computing: A Compendium of Select, Pivotal Pioneers*,
DOI 10.1007/978-1-4471-5340-5_21, © Springer-Verlag London 2013

Fig. 21.1 George Devol

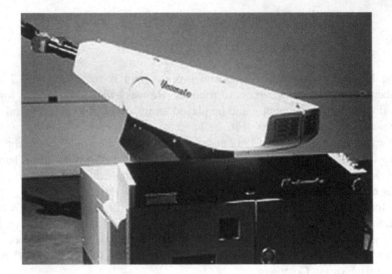

Fig. 21.2 Unimate in the 1960s

They initially developed a material handling robot and robots for welding and other applications followed. They installed the first industrial robot (Unimate) on a General Motors (GM) assembly line in 1961. This robot was used to lift hot pieces of metal from a die-casting machine and to stack them. The story of Unimation Inc. is told in [Mun:10].

These robots were very successful and reliable and saved General Motors money by replacing staff with machines. Other automobile companies followed GM in purchasing Unimate robots, and the robot industry continues to play a major role in the automobile sector.

A Unimate robot appeared on *The Tonight Show* hosted by Johnny Carson in 1966, and the robot poured a beer and sank a golf putt. Devol has received various awards for his contributions to robotics. His Unimate was named by *Popular Mechanics* magazine in 2005 as one of the top 50 inventions of the past 50 years. He was inducted into the National Inventors Hall of Fame in 2011. He died aged 99 in Connecticut 2011.

21.1 Robotics

The first use of the term "robot" was by the Czech playwright Karel Capek in his play *Rossum's Universal Robots* performed in 1921. The word *robot* is a Czech word for forced labour, and the play explores whether it is ethical to exploit artificial workers in a factory and how robots should respond to exploitation. Capek's robots looked and acted like humans and were created by chemical means. He rejected the idea that machines created from metal could think or feel.

Asimov wrote several stories about robots in the 1940s including the story of a robotherapist. He predicted the rise of a major robot industry and introduced a set of rules (or laws) of good robot behaviour (Table 21.1).

Robots have been very effective at doing clearly defined repetitive tasks, and there are many sophisticated robots in the workplace today. These are industrial manipulators that are essentially computer controlled "arms and hands". The term *robot* is defined by the Robot Institute of America as:

Definition 21.1 (Robots) A re-programmable, multifunctional manipulator designed to move material, parts, tools, or specialized devices through various programmed motions for the performance of a variety of tasks.

Robots can also improve the quality of life for workers as they can free human workers from performing dangerous or repetitive jobs. They provide consistently high-quality products and can work tirelessly 24 h a day. This helps to reduce the costs of manufactured goods, thereby benefiting consumers.

Table 21.1 Asimov's laws of robotics

Law	Description
Law zero	A robot may not injure humanity, or, through inaction, allow humanity to come to harm
Law one	A robot may not injure a human being, or, through inaction, allow a human being to come to harm, unless this would violate a higher-order law
Law two	A robot must obey orders given it by human beings, except where such orders would conflict with a higher-order law
Law three	A robot must protect its own existence as long as such protection does not conflict with a higher-order law

Chapter 22
Larry Ellison

Larry Ellison is an American entrepreneur and the co-founder and chief executive officer (CEO) of Oracle Corporation. This is one of the leading enterprise software companies in the world and is well-known for its database management products. It also produces enterprise resource planning (ERP) and customer relationship management (CRM) software (Fig. 22.1).

He was born in New York in 1944 and was adopted by his aunt in Chicago when he was just 9 months old. He attended the University of Illinois, but left at the end of his second year due to the death of his adoptive mother. He moved to California in 1964 and worked briefly for the Amdahl Corporation in the 1970s.

He moved to Ampex Corporation, a private American electronics company, and one of his early work assignments was a database project for the CIA. This project was code-named *Oracle*, and he was later to name his company and its flagship product "Oracle". He became familiar with Codd's theoretical work on relational databases (discussed in an earlier chapter) and with IBM's System R research project on relational databases. He founded Software Development Laboratories (SDL) in 1977, and the company was renamed to Relational Software Inc. in 1979. It was later renamed to *Oracle* after the company's main product.

IBM dominated the mainframe database market in the late 1970s with products such as DB2 and SQL/DS. However, it delayed entry into the market for relational databases on UNIX and Windows operating systems, and this allowed companies such as Sybase, Oracle and Informix to target this market segment and to become the dominant players in these markets.

Microsoft took over Sybase in 1993, and Sybase's existing Microsoft Windows database product became known as "SQL Server". This product was to become a major competitor to Oracle from the late 1990s. IBM took over Informix in 2001 to complement its DB2 product database product. IBM and Microsoft are major competitors for Oracle on the UNIX, Linux and Microsoft Windows operating systems.

G. O'Regan, *Giants of Computing: A Compendium of Select, Pivotal Pioneers*, DOI 10.1007/978-1-4471-5340-5_22, © Springer-Verlag London 2013

Fig. 22.1 Larry Ellison
on stage

22.1   Oracle Corporation

Larry Ellison, Bob Miner and Ed Oates founded Software Development Laboratories in 1977. Ellison became the chief executive of the company and was responsible for sales and marketing, while Miner was responsible for software development. The company changed its name to Oracle Systems in 1983. Ellison saw an opportunity to exploit and commercialize the relational database technology developed by Codd before IBM.

There were no commercial relational databases in the world in the late 1970s, and the launch of Oracle's database changed business processes and computing. Its database product was called Oracle, and the release of Oracle V.2 in 1979 was an important milestone in the history of computing. It was the first commercial SQL relational database management system.

Oracle is a world leader in relational database technology and its products play a key role in business computing. An Oracle database consists of a collection of data managed by an Oracle database management system. It is the main standard for database technology and is used by companies throughout the world. It has become one of the largest software companies in the world with over 100,000 employees.

It is also involved in the enterprise application market and has acquired companies such as PeopleSoft and Siebel. It has become a competitor to SAP in the enterprise resource planning (ERP) and customer relationship management (CRM) market. This led to a legal dispute between Oracle and SAP over allegations that an acquired subsidiary of SAP, Tomorrow Now, had illegally downloaded Oracle software.

22.2   Oracle Database

An Oracle database is a collection of data treated as a unit, and a database is used to store and retrieve related information. It is placed on a server, and the database server manages a large amount of data in a multi-user environment. It allows concurrent access to the data, and the database management system prevents unauthorized access to the database. It also provides a smooth recovery of database information in the case of an outage or any other disruptive event.

Every Oracle database consists of one or more physical data files, which contain all of the database data, and a control file that contains entries that specify the physical structure of the database.

An Oracle database includes logical storage structures that enable the database to have control of disc space use. A schema is a collection of database objects, and the schema objects are the logical structures that directly refer to the database's data. They include structures such as tables, views and indexes.

Tables are the basic unit of data storage in an Oracle database, and each table has several rows and columns. An index is an optional structure associated with a table, and it is used to enhance the performance of data retrieval. The index provides an access path to the table data. A view is the customized presentation of data from one or more tables. It does not contain actual data and derives the data from the actual tables on which it is based.

Each Oracle database has a data dictionary, which stores information about the logical and physical structure of the database. The data dictionary is created when the database is created and is updated automatically by the Oracle database to ensure that it accurately reflects the status of the database at all times.

An Oracle database uses memory structures and processes to manage and access the database. These include server processes, background processes and user processes.

A database administrator (DBA) is responsible for setting up the Oracle database server and application tools. This role is concerned with allocating system storage and planning future storage requirements for the database management system. The DBA will create appropriate storage structures to meet the needs of application developers who are designing a new application. The access to the database will be monitored and controlled, and the performance of the database monitored and optimized. The DBA will plan backups and recovery of database information.

Chapter 23
Don Estridge

Philip Donald Estridge is known as the *father of the IBM PC*. He led the development of the original IBM personal computer, which revolutionized the computer industry. It led to millions of computers in business offices and homes, and today, personal computers and laptops are used extensively around the world (Fig. 23.1).

He was born in Florida in 1937, and he studied electrical engineering at the University of Florida. He was awarded a bachelor's degree in 1959. He then worked at the US Army designing a radar system using a computer, and he joined IBM as an engineer in New York later that year. He spent most of his early career with IBM Federal Systems Division and worked on the SAGE system and later at NASA's Goddard Space Flight Centre on computer systems for the Apollo mission to the moon.

He became manager of the IBM Entry Level Systems in 1980, and his goal was to develop a low-cost personal computer to compete against product offerings from Apple, Commodore and other vendors. He realized that he would need to use third-party hardware and software in order to deliver the IBM PC in a timely manner and at an affordable price for consumers. The reliance on third-party suppliers was a major shift in IBM's business practices. Estridge published the specification of the IBM PC, which led to a new industry of IBM compatibles and to new hardware businesses who took advantage of the expansion slots in the IBM PC.

He was appointed president of IBM Entry Systems Division in 1983 and was promoted to vice president of IBM Manufacturing in 1984. Over one million personal computers were sold by 1985.

Estridge and his wife were killed in a Delta airlines air crash at Dallas International Airport in 1985. The accident was caused by wind shear while landing at the airport. He was just 48 years old.

G. O'Regan, *Giants of Computing: A Compendium of Select, Pivotal Pioneers*,
DOI 10.1007/978-1-4471-5340-5_23, © Springer-Verlag London 2013

Fig. 23.1 Don Estridge
(Courtesy of IBM archives)

23.1   IBM Personal PC

IBM introduced the IBM personal computer (or PC) in 1981 as a machine to be used by small businesses and personal users in the home. Its strategy at the time was to get into the home computer market that was then dominated by Commodore, Atari and Apple. The IBM PC was priced at $1,565, and it was affordable to personal computer users. It offered 16 kB of memory (expandable to 256 kB), a floppy disc and a monitor. It was an immediate success and became the industry standard.

IBM's goal was to get the personal computer to the market quickly, and a small team of 12 people led by Don Estridge was assembled. The chief designer was Lewis Eggebrecht. The goal was to design and develop the IBM PC within 1 year, and as time to market was a key driver, they built the machine with *off-the-shelf* parts from a number of equipment manufacturers. They had intended using the IBM 801 processor developed at the IBM Research Centre in Yorktown Heights, but decided instead to use the Intel 8088 microprocessor, which was inferior to the IBM 801. They chose the PC-DOS operating system from Microsoft rather than developing their own operating system (Fig. 23.2).

The unique IBM elements in the personal computer were limited to the system unit and keyboard. The team decided on an open architecture so that other manufacturers could produce and sell peripheral components and software without purchasing a licence. They published the *IBM PC Technical Reference Manual*, which included the complete circuit schematics, the IBM ROM BIOS source code and other engineering and programming information.

Estridge and the development team came up with a prototype machine within 4 months, and the completed personal computer was available within the 1-year development goal. The IBM personal computer overtook Apple as the bestselling personal computer in 1983.

The open architecture led to a new industry of *IBM-compatible* computers, which had all of the essential features of the IBM PC, but were cheaper. The terms of the licensing of PC-DOS operating system gave Microsoft the rights to the MS-DOS

Fig. 23.2 IBM personal computer (Courtesy of IBM archives)

operating system used on the IBM compatibles, and this decision led inexorably to the rise of the Microsoft Corporation. The IBM Personal Computer XT was introduced in 1983. This model had more memory, a dual-sided diskette drive and a high-performance fixed-disc drive. It was followed by the introduction of the Personal Computer AT introduced in 1984.

The development of the IBM PC meant that computers were now affordable to ordinary users, and this led to a huge consumer market for personal computers and software. It led to the development of business software such as spreadsheets and accountancy packages, banking packages, programmer developer tools such as compilers and specialized editors and computer games.

The introduction of the personal computer represented a paradigm shift in computing, and it placed the power of the computer directly in the hands of millions of people.

IBM had traditionally produced all of the components for its machines. However, it outsourced the production of components to other companies for the IBM PC. This proved to be a *major error* as they outsourced the production of the processor chip to a company called Intel,[1] and the development of the disc operating system (DOS) was outsourced to a small company called Microsoft.[2] Intel and Microsoft would later become technology giants.

---

[1] Intel was founded by Bob Noyce and Gordon Moore in 1968.
[2] Microsoft was founded by Bill Gates and Paul Allen in 1975.

Chapter 24
Michael Fagan

Michael Fagan is the CEO of Michael Fagan Associates, and the creator of the Fagan Inspection and Defect-Free Process. He created the Fagan inspection process while he was a Development Manager with IBM in the 1970s. This process helps organizations to improve software quality, to reduce cycle time, to reduce costs and to improve productivity (Fig. 24.1).

He attended King's School, a grammar school located in Lincolnshire, England. He was one of the founder members of the IBM Quality Institute in 1981. He was a visiting professor at the University of Maryland from 1983 to 1985 and a senior technical staff member conducting research in software engineering at IBM Research in Yorktown Heights, New York, from 1985 to 1989. Fagan received an individual corporate achievement award from IBM for creating the Fagan inspection process and for promoting its use throughout IBM.

He founded Michael Fagan Associates in 1989 to teach others how to use this inspection process effectively and to improve software quality in organizations throughout the world.

24.1 Fagan Inspection Process

Software inspections play an important role in identifying faults in the software development life cycle and in building quality into the software product. There is clear evidence that the cost of correction of a defect increases the later that it is detected in the development cycle. Consequently, there is an economic argument to identify defects as early as possible, as it is more cost-effective to build quality into the software product rather than adding quality later in the development life cycle.

The Fagan inspection process was developed by Michael Fagan at IBM in the early 1970s [Fag:76], and the process identifies and removes defects in the work products. It stipulates that requirement documents, design documents, source code and test plans all be formally inspected by experts independent of the author, and that the deliverable be examined from different viewpoints. The quality of the

Fig. 24.1 Michael Fagan

software product is only as good as the quality at the end of each phase, and software inspections assist in ensuring that quality has been built into each phase and in ensuring that the final product is fit for purpose.

The seven-step process includes planning, overview, preparation, inspection, process improvement, rework and follow-up activity. Its objectives are to identify and remove errors in the work products and to identify any systemic defects in the processes used to create the work products.[1]

There are various roles defined in the inspection process, including the *moderator*, who chairs the inspection; the *reader*, who paraphrasing the particular deliverable and gives an independent viewpoint; the *author*, who is the creator of the deliverable; and the *tester*, who is concerned with the testing viewpoint. The inspection process will consider whether a design is correct with respect to the requirements and whether the source code is correct with respect to the design. There seven stages in the process are summarized in Table 24.1.

There is tangible evidence that software inspections have positive impacts on productivity, quality, time to market and customer satisfaction. For example, IBM Houston employed software inspections for the Space Shuttle missions, and the results showed that 85 % of the defects were found by inspections with 15 % were found by testing. There were no defects found in the space missions. This project includes about two million lines of computer software. IBM, North Harbour in the UK, quoted a 9 % increase in productivity with 93 % of defects found by software inspections.

Software inspections also play an important role in educating and training new employees about the product and the standards and procedures to be followed. Sharing knowledge reduces dependencies on key employees. Higher-quality software leads to improved productivity, as less time is devoted to reworking the defective product.

---

[1]A defective process may lead to downstream defects in the work products.

Table 24.1 Fagan inspections

Step	Description
Planning	This includes identifying the inspectors and their roles, providing copies of the inspection material to the participants and booking rooms for the inspection
Overview	The author provides an overview of the deliverable to the inspectors (optional step)
Prepare	All inspectors prepare for the inspection and the role that they will perform
Inspection	The actual inspection takes place and the emphasis is on finding major errors (not solutions)
Process improvement	This part is concerned with continuous improvement of the development process and the inspection process
Rework	The defects identified during the inspection are corrected, and any items requiring investigation are resolved
Follow-up activity	The moderator verifies that the author has made the agreed-upon corrections and completed any required investigations

The cost of correction of a defect increases the later it is identified in the life cycle. Boehm [Boe:81] states that a requirement's defect identified by the customer is over 40 times more expensive to correct than if it were detected in the requirement's phase. The cost of a requirement's defect detected at the customer site includes the cost of correcting the requirements and the cost of design, coding, unit testing, system testing, and regression testing. It is more economical to detect and fix the defect in phase.

The inspection process will consider whether a design is correct with respect to the requirements and whether the source code is correct with respect to the design. The errors identified in an inspection are typically classified into various types, and mature organizations record the inspection data in a database for further analysis. Measurement allows the effectiveness of the organization in identifying errors in phase and detecting defects out of phase to be determined and improved.

Tom Gilb has proposed an alternate inspection methodology [Glb:94], and both Fagan and Gilb inspections focus on the management aspects of software inspections. These approaches have yielded good results and have led to higher quality, improved productivity and shortened development time. The Fagan inspection process is described in more detail in [ORg:02].

Software inspections have been criticized by Parnas and others on the grounds that they lack the rigour associated with mathematical reasoning. That is, they do not provide a rigorous mechanism to ensure that all cases have been considered, or that a particular case is correct. Parnas has defined a mathematical approach to software inspections for the safety critical field [ORg:06].

Chapter 25
Tommy Flowers

Tommy Flowers was a British engineer who made important contributions to breaking the Lorenz codes during the Second World War. He led the team that designed and built Colossus, which was one of the earliest electronic computers. The machine was designed to decode the top-level encrypted German military communication sent by German High Command to its commanders in the field. This provided British and American Intelligence with information on German military plans around the D-Day invasion and later battles and helped to ensure the success of the Normandy landings and the ultimate defeat of Nazi Germany (Fig. 25.1).

He was born in East London in 1905 and attended evening classes in electrical engineering at the University of London, while serving an apprenticeship in mechanical engineering at the Royal Arsenal in Woolwich. He was awarded a degree in electrical engineering from the university and obtained a position with the telecommunications branch of the General Post Office in 1926. He moved to the research station at Dollis Hill in 1930 and investigated the use of electronics for telephone exchange. He was convinced at an early stage that an all-electronic system was possible.

Flowers became involved with the code-breaking work taking place at Bletchley Park near Milton Keynes during the Second World War. Alan Turing and others had cracked the German Enigma codes by building a machine known as the Bombe. This machine employed a crib to deduce the settings of the Enigma machine for that day. Turing introduced Flowers to Max Newman who was leading British efforts to break a German cipher generated by the Lorenz SZ42 machine. The British referred to this teletypewriter-coding machine as *Tunny*, and it was considerably more complex than the Enigma machine.

Their existing approach to deciphering the Lorenz codes was with the Heath Robinson machine (a slow and unreliable machine). Flowers proposed an alternate solution involving the use of an electronic machine in 1943. This machine was called Colossus and it employed 1,800 thermionic valves. The management at Bletchley Park were sceptical, but encouraged him to continue with his work.

Flowers and others at the Post Office Research Centre built the machine in 11 months, and its successor, the Mark II Colossus, contained 2,400 valves and

G. O'Regan, *Giants of Computing: A Compendium of Select, Pivotal Pioneers*,
DOI 10.1007/978-1-4471-5340-5_25, © Springer-Verlag London 2013

Fig. 25.1 Tommy Flowers

commenced operations on June 1, 1944. It was a large bulky machine and took up the space of a small room and weighed a tonne.

It provided vital information for the Normandy landings and confirmed that Hitler had been successfully misled by Allied disinformation into believing that the Normandy landings were to be a diversionary tactic. Further, it confirmed that no additional German troops were to be moved there. The Colossus Mark II machine helped the British to monitor the German reaction to their deception tactics.

Flowers returned to Dollis Hill after the war and became the head of the switching group. He received several awards in recognition of his contributions. He received a Member of the order of the British Empire (MBE) in 1943 and received the Martlesham Medal from the Post Office in 1980. He died in 1998 aged 92 years.

25.1 Colossus

Flowers and others designed and built the original Colossus machine at the Post Office Research Station at Dollis Hill in London. The machine was used to find possible key combinations for the Lorenz machines rather than decrypting an intercepted message in its entirety. The Lorenz machine was based on the *Vernam cipher*.

It compared two data streams to identify possible key settings for the Lorenz machine. The first data stream was the encrypted message, and it was read at high speed from a paper tape. The second stream was generated internally and was an electronic simulation of the Lorenz machine at various trial settings. If the matched count for a setting was above a certain threshold, it would be sent as output to an electric typewriter.

The Lorenz codes were a more complex cipher than the Enigma codes, and they were used in the transmission of important messages between the German High Command in Berlin and the military commanders in the field. The Lorenz SZ 40/42 machine performed the encryption. The Bletchley Park code breakers called the

Fig. 25.2 Colossus Mark II (Photo courtesy of the UK government)

machine *Tunny* and the coded messages *Fish*. The code-breaking work involved carrying out complex statistical analyses on the intercepted messages.

The Colossus Mark I machine was specifically designed for code breaking rather than as a general-purpose computer. It was semi-programmable and helped in deciphering messages encrypted using the Lorenz machine. A prototype was available in 1943 and a working version was available in early 1944 at Bletchley Park. The Colossus Mark II was introduced just prior to the Normandy landings in June 1944 (Fig. 25.2).

The Colossus Mark I used 15 kW of power and could process 5,000 characters of paper tape per second. It enabled a large amount of mathematical work to be done in hours rather than in weeks. There were ten Colossi machines working at Bletchley Park by the end of the war. A replica of the Colossus was rebuilt by a team of volunteers led by Tony Sale from 1993 to 1996 and is at Bletchley Park museum.

The contribution of Bletchley Park to the cracking of the German Enigma and Lorenz codes and to the development of computing remained clouded in secrecy until recent times. The museum at Bletchley Park provides insight to the important contribution made by this organization during the Second World War.

25.2 Vernam Cipher

Gilbert Vernam was an AT&T research engineer who invented the Vernam cipher. He invented a stream cipher in 1917 and co-invented the *one-time pad* cipher with Joseph Mauborgne. Claude Shannon later showed that if the one-time pad cipher is correctly implemented, then it is theoretically unbreakable, and that any unbreakable system must have essentially the same characteristics as the one-time pad.

The stream cipher proposed by Vernam involved using a previously prepared key kept on a paper tape to combine character by character with the plaintext message to form the ciphertext. The deciphering of the message involved using the same key combining character by character with the ciphertext to form the plaintext.

His encryption system used conventional telegraphy practice with the paper tape of the plaintext combined with the paper tape of the key. The intention was that each key tape would be unique, but there were practical problems with generating and distributing such tapes. This problem was solved in the 1920s with the invention of rotor cipher machines to produce a key stream to act instead of a tape. The Lorenz SZ40/42 was one of these.

The encryption and decryption is defined by addition modulo 2, and it is symmetric with the same key employed for both encryption and decryption. It is defined by

$$\text{Ciphertext} = \text{Plaintext} \oplus \text{Key}$$

$$\text{Plaintext} = \text{Ciphertext} \oplus \text{Key}$$

where the $\oplus$ symbol means that a logical or (XOR) operation is performed.

If the key stream is truly random and is used only once, then this is effectively a one-time pad. *However, if the key stream is used for two messages, then the effect of the key stream can be eliminated*, and it is possible for cryptanalysts to derive the message by linguistic cryptanalytical techniques.

$$\text{Ciphertext}_1 \oplus \text{Ciphertext}_2 = \text{Plaintext}_1 \oplus \text{Plaintext}_2$$

It was operator error of this sort that enabled the British to crack the Lorenz codes.

Chapter 26
Robert Floyd

Robert Floyd was an American computer scientist who made important contributions to the theory of parsing and early compilers, to the semantics of programming languages and to the development of methodologies for the creation of efficient and reliable software (Fig. 26.1).

He was born in New York in 1936 and studied at the University of Chicago. He obtained a bachelor's degree in liberal arts in 1953 and a second bachelor's degree in physics in 1958.

He began publishing in computer journals from 1958 and became a computer operator and senior programmer in the early 1960s. He became a senior project scientist at Computer Associates (a software company that developed compilers) in Massachusetts in 1962, and he became interested in the syntax of programming languages and parsing [Flo:63, Flo:64].

He became an associate professor at Carnegie Mellow University in 1963 and a full professor of computer science at Sanford University in 1969. He did pioneering work on software engineering from the 1960s and made important contributions to the theory of parsing, the semantics of programming languages and program verification.

Mathematics and computer science were regarded as two completely separate disciplines back in the 1960s, and software development was based on the assumption that the completed code would always contain defects. It was therefore better and more productive to write the code as quickly as possible and to then perform debugging to find the defects in the software. Programmers then corrected the defects, made patches, retested and found more defects. The process continued until they could no longer find defects. Of course, there was always the danger that defects remained in the code that could give rise to software failures.

Floyd challenged this assumption and he believed that there was a way to construct a rigorous proof of the correctness of the programs using mathematics. He did important pioneering work in program verification and showed that mathematics could be employed to prove the correctness of programs. He introduced the concept of invariant assertions which provided a way to verify the correctness of programs.

G. O'Regan, *Giants of Computing: A Compendium of Select, Pivotal Pioneers*,
DOI 10.1007/978-1-4471-5340-5_26, © Springer-Verlag London 2013

Fig. 26.1 Robert Floyd

Fig. 26.2 Branch assertions
in flow charts

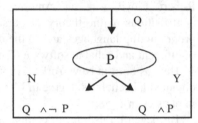

His pioneering 1967 paper "Assigning meanings to programs" [Flo:67] founded the new field of program verification. His key insight was that a computer program is essentially a sequence of logical assertions, with each assertion true whenever control passes to it. Statements appear between the assertions and the logical assertions before and after the program fragment describes the effect of the individual program statements or branches of a flow chart. His approach was to influence C.A.R. Hoare's work on Hoare's pre- and postconditions and led to Hoare logic.

Flow charts were employed in the 1960s to explain the sequence of basic steps for computer programs. Floyd's insight was to build upon flow charts and to apply an invariant assertion to each branch in the flow chart. These assertions state the essential relations that exist between the variables at that point in the flow chart. An example relation is "$R = Z > 0, X = 1, Y = 0$".

He devised a general flow chart language which consists essentially of boxes linked by the flow of control arrows. Consider the assertion Q that is true on entry to a branch where the condition at the branch is P. Then, the assertion on exit from the branch is $Q \land \neg P$ if P is false and $Q \land P$ otherwise (Fig. 26.2).

The use of assertions may be employed in an assignment statement. Suppose x represents a variable and v represents a vector consisting of all the variables in the program. Suppose $f(x,v)$ represents a function or expression of x and the other program variables represented by the vector v. Suppose the assertion $S(f(x,v), v)$

Fig. 26.3 Assignment
assertions in flow charts

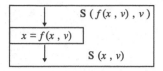

is true before the assignment $x = f(x,v)$. Then the assertion $S(x,v)$ is true after the assignment. This is given by (Fig. 26.3):

Floyd used flow chart symbols to represent entry and exit to the flow chart. He used entry and exit assertions to describe the program's entry and exit conditions.

Floyd's technique showed how a computer program is a sequence of logical assertions. Each assertion is true whenever control passes to it and statements appear between the assertions. The initial assertion states the conditions that must be true for execution of the program to take place, and the exit assertion essentially describes what must be true when the program terminates.

He recognized that if it can be shown that the assertion immediately following each step is a consequence of the assertion immediately preceding it, then the assertion at the end of the program will be true, provided the appropriate assertion was true at the beginning of the program.

He received the ACM Turing Award in 1978 for his influence on the development of methodologies for the creation of efficient and reliable software and for helping to found several important subfields of computer science such as the theory of parsing and semantics of programming languages, program verification and analysis of algorithms.

He received the IEEE Computer Pioneer Award in 1992 for his work on early compilers. A compiler is a program that checks the syntax correctness of a program and converts it into the equivalent machine code. The compiler preserves the semantics of the language.

Floyd worked closely with Donald Knuth and was the major reviewer of Knuth's influential *The Art of Computer Programming*. He was the co-author, with Richard Beigel, of the textbook *The Language of Machines*: *An Introduction to Computability and Formal Languages*, which appeared in 1994.

Floyd and others at Stanford played an active role in securing the release of Fernando Flores from prison in Chile, where he was held by the Pinochet's military regime. Floyd died of a neurodegenerative illness in 2001.

Chapter 27
Bill Gates

Bill Gates is an American entrepreneur, computer programmer and philanthropist. He is the former chief executive officer (CEO) of Microsoft Corporation, a company that he co-founded with Paul Allen. He is currently the chairman of the board of the company (Fig. 27.1).

He was born in Seattle, Washington, in 1955 and became interested in computers and programming while at Lakeside school. He wrote his earliest program in BASIC on a General Electric computer via a Teletype Model 33 ASR terminal, and he spent time at the Computer Center Corporation studying source code on their PDP-10 system including FORTRAN, LISP and machine languages. Gates, Allen and two other Lakeside students were hired by Information Sciences, Inc., in 1971 to work on a payroll system in COBOL.

He enrolled at Harvard College in 1973 and met Steve Ballmer the same year.[1] Gates spent a lot of time on the college's computers at Harvard and remained in touch with Paul Allen. The Altair 8800 (the first commercially successful home computer) was released in 1975 and appeared on the cover of the January 1975 edition of the magazine *Popular Electronics*. Gates and Allen spotted the announcement and began writing software for the machine.

The Altair 8800 was developed at MITS (Micro Instrumentation Telemetry Systems), and it was initially purchased and assembled by computer hobbyists. Gates and Allen spent a short time working with MITS in 1975 and founded Microsoft later that year. The company had 11 employees by 1978, and their first product was the Altair BASIC interpreter.

IBM approached Microsoft in 1980 and contracted Microsoft to write a BASIC interpreter for the new IBM personal computer. IBM's discussions with Digital Research on the licensing of their CP/M operating system as the operating system for the new personal computer were unsuccessful,[2] and Gates offered to provide IBM with an operating system (called PC-DOS) that was based on a CP/M clone

---

[1]Ballmer later joined Microsoft and when Gates stepped down as CEO he was replaced by Ballmer.
[2]This is discussed further in the chapter on Gary Kildall.

G. O'Regan, *Giants of Computing: A Compendium of Select, Pivotal Pioneers*,
DOI 10.1007/978-1-4471-5340-5_27, © Springer-Verlag London 2013

Fig. 27.1 Bill Gates
delivering a keynote speech
in 2004

called QDOS. The terms of the licensing agreement with IBM allowed Microsoft to
offer their own version of the operating system (called MS-DOS) to other hardware
vendors who cloned IBM's system. This led to major sales of MS-DOS (as many
hardware vendors took advantage of the open specification of the personal computer
to produce clones of the IBM personal computer), and Microsoft's MS-DOS became
the dominant operating system for personal computers. This led to the rise of the
Microsoft Corporation.

The company has since developed a suite of products for personal computers
and laptops, including Microsoft Word, Microsoft Excel, Microsoft PowerPoint,
Microsoft Office, Microsoft Windows, and Microsoft XP.

Gates and his wife established Bill and Melinda Gates Foundation in 2000, and
this charitable foundation is active in philanthropic activities around the world.

Gates has received various awards in recognition of his achievements. *TIME*
magazine has named him as one of the 100 people who have most influenced the
twentieth century. He and his wife were listed as persons of the year by *TIME*
magazine for their humanitarian efforts.

27.1 Microsoft

Gates and Allen formed Microsoft in 1975, and Steve Ballmer joined the company
in 1980. Its first real success was with the disc operating system (DOS) for personal
computers. IBM had intended awarding the contract to develop the operating system
for the personal computer to Digital Research (the company that had developed the
CP/M operating system). However, negotiations between IBM and Digital Research
broke down in 1981, and IBM awarded the contract to Microsoft to produce the PC-
DOS for the personal computer.

Microsoft purchased a CP/M clone called QDOS and enhanced it for the IBM
personal computer. IBM renamed the new operating system to PC-DOS, and
Microsoft created its own version of the operating system called MS-DOS. The

terms of the contract allowed Microsoft to have control of its own QDOS derivative, and this allowed Microsoft to offer it to other hardware vendors who had cloned the IBM personal computer. This proved to be a major mistake by IBM, as MS-DOS became popular in Europe, Japan and South America. The flood of PC clones to the market allowed Microsoft to gain major market share with effective marketing of MS-DOS to the various manufacturers of the cloned PCs. This led to Microsoft becoming a major player in the personal computer market.

The company developed application software to run on its operating system. It released its first version of Microsoft Word in 1983, and this would later become the world's most popular word processing package. The Microsoft *Office* suite of products was introduced in 1989, and these include Microsoft *Word*, *Excel* and *PowerPoint*. Microsoft's office suite gradually became the dominant office suite with a far greater market share than its competitors such as WordPerfect and Lotus 1-2-3.

Microsoft released its first version of Microsoft Windows in 1985, and this product was a graphical extension of its MS-DOS. Microsoft and IBM commenced work in 1985 on a new operating system called Operating System 2 (OS/2) for the IBM PS/2 personal computer; however, this operating system never became popular.

It introduced Windows 3.0, in 1990, and this operating system included a friendly graphical user interface. Windows (and its successors) became the dominant operating system for the IBM PC. Microsoft released its Windows 3.1 operating system and its Access database software in 1992. Windows 95 was released in 1995, Windows NT in 1996, Windows 2000 in 2000, Windows XP in 2001, Windows 7 in 2009 and Windows 8 in 2012.

27.1.1 Microsoft Windows and Apple GUI

Apple Computers took a copyright infringement lawsuit against Microsoft in 1988 to prevent Microsoft from using its GUI elements in the Windows operating system. The legal arguments lasted 5 years and the final ruling in 1993 was in favour of Microsoft. Apple had claimed that the look and feel of the Macintosh operating system was protected by copyright including 189 GUI elements. However, the judge found that 179 of these had already been licensed to Microsoft (as part of the Windows 1.0 agreement), and that most of the 10 other GUI elements were not copyrightable.

This legal case generated a lot of interest as some observers considered Apple to be the villain, as they were using legal means to dominate the GUI market and restrict the use of an idea that was of benefit to the wider community. Others considered Microsoft to be the villain with their theft of Apple's work, and their argument was that if Microsoft succeeded, a precedent would be set in allowing larger companies to steal the core concepts of any software developer's work.

The court's judgement seemed to invalidate the copyrighting of the *look and feel* of an application. However, the judgement was based more on contract law rather than copyright law, as Microsoft and Apple had previously entered into a contract with respect to licensing of Apple's icons on Windows 1.0. Apple had not acquired a software patent to protect its intellectual idea of the look and feel of its Macintosh operating system.

27.1.2   The Browser Wars

Microsoft was initially slow to respond to the rise of the Internet. It developed Microsoft Network (MSN) to compete directly against America On-Line (AOL). It developed some key Internet technologies such as ActiveX, VBScript and Jscript, and it released a new version of Microsoft SQL Server with built-in support for Internet applications. It released Internet Explorer 4.0 (its Internet browser[3]) in 1997, and Internet Explorer was released as part of Windows operating system.

This was the beginning of Microsoft's dominance of the browser market. Netscape[4] had dominated the market, but as Internet Explorer 4.0 (and its successors) was provided as a standard part of the Windows operating system (and also on Apple computers), this inevitably led to the replacement of Netscape by Internet Explorer.

Netscape launched a legal case against Microsoft alleging that Microsoft was engaged in anticompetitive practices by including the Internet Explorer browser in the Windows operating system, and that Microsoft had violated an agreement signed in 1994. The leaking of internal Microsoft company memos caused a great deal of controversy in 1998, as these documents went into detail of the threat that open source software posed to Microsoft and mentioned possible legal action against Linux and other open source software.

The Federal Trade Commission and Department of Justice in the United States investigated Microsoft on various antitrust allegations in the early 1990s. The 1991–1994 investigations by the Federal Trade Commission ended with no lawsuits. The 1994 investigation by the Department of Justice ended in 1995 with a consent decree. The 1995 decree imposed restrictions on Microsoft and prohibited bundling of certain products. The Department of Justice alleged in 1997 that Microsoft violated the 1995 agreement by bundling Internet Explorer with its Windows operating system and requiring manufacturers to distribute Internet Explorer with Windows 95.

The Court of Appeals rejected this violation of the consent decree in 1998 and stated that the 1995 consent decree did not apply to Windows 98 which was shipped with an integrated Internet Explorer as part of the operating system. The Department

---

[3]A browser is a software package with a graphical user interface that allows a user to view the World Wide Web.

[4]Netscape was founded by Jim Clark and Mark Andreessen in 1994. Jim Barksdale became the president and CEO of Netscape in 1995.

of Justice then filed a major antitrust suit against Microsoft in 1998 and argued that Microsoft's bundling of Internet Explorer and its attempts to eliminate Netscape as a competitor in the browser market were much more than adding functionality to its operating system. It alleged that Microsoft added browser functionality to Windows and marginalized Netscape because Netscape posed a potential threat to the Windows operating system. It alleged that Microsoft feared that since Netscape could run on several operating systems, this could erode the power of Windows as applications could be written on top of Netscape.

In other words, the Department of Justice alleged that Microsoft gave away Internet Explorer and bundled it with its operating system to prevent Netscape becoming a platform that would compete with Microsoft. That is, Microsoft's actions were a defensive move to protect its Windows monopoly. The legal action concluded in mid-2000, and the judgement called the company an *abusive monopoly*. The judgement stated that the company should be split into two parts. However, this ruling was subsequently overturned on appeal.

27.2   Bill and Melinda Gates Foundation

Gates sold some of his Microsoft stock in 1994 and founded the William H. Gates Foundation as an organization dedicated to philanthropy. He married Melinda French in 1994, and they set up the Bill and Melinda Gates Foundation in 2000. The foundation is tackling poverty and poor health in developing countries around the world. The Foundation believes that innovation can play a key role in improving the world, and it supports innovation in farming and in areas of global development and global health. Innovation in farming includes identifying new techniques to help farmers to grow more food and earn more money (e.g. knowledge about managing soil and irrigation and in preventing crop diseases). The lack of adequate nutrition leads to basic health problems.

Most of the Foundation's resources go into global health issues, and it has been active in identifying new ways of tackling deadly diseases. It has invested in vaccines for children to protect them from deadly diseases, and it is active in the eradication of polio and in supporting the prevention of HIV and AIDS. It has also supported new methods to help teachers and students in the classroom. It has invested billions of dollars in many projects around the world.

Chapter 28
James Gosling

James Gosling is a Canadian computer scientist who is regarded as the father of the Java programming language. Gosling, Patrick Naughton and others at Sun Microsystems designed the language and implemented the original compiler and virtual machine. The language was originally called "Oak", but it was renamed to "Java" in 1995. It is a popular language for developing application software and may be deployed in a cross-platform computing environment (Fig. 28.1).

He was born in Alberta, Canada, in 1955 and obtained a bachelor's degree in computer science from the University of Calgary in 1977. He moved to the United States to pursue postgraduate studies and earned a Ph.D. in computer science from Carnegie Mellon University in 1983.

He wrote several compilers, email systems and a version of Emacs (a family of text editors) during his postgraduate years at Carnegie Mellon. He joined Sun Microsystems in 1984 and developed the Java programming language in 1994. He is the author of several books on Java. He remained with Sun Microsystems until 2010, when it was taken over by the Oracle Corporation.

He resigned from Oracle and is currently an employee of Liquid Robotics. This company specializes in ocean data services, and it has developed an unmanned robotic marine vehicle called Wave Glider. This marine robot uses wave energy for propulsion and monitors and collects ocean data.

He has received various awards for his contributions to the computing field, including the 2007 award of an Officer of the Order of Canada.

28.1 Java

The rise of the Internet and the World Wide Web has led to fundamental changes in the computing field. The existing paradigm was dominated by stand-alone personal computers, whereas today nearly all computers are connected to the Internet. The latter was initially used to share files and information, whereas today it is a vast distributed computing space. Java has played a key role in transforming the Internet,

G. O'Regan, *Giants of Computing: A Compendium of Select, Pivotal Pioneers*,   129
DOI 10.1007/978-1-4471-5340-5_28, © Springer-Verlag London 2013

Fig. 28.1 James Gosling

and it has fundamentally changed the way in which people program. *It supports application programming in a distributed computing environment.*

Java is a collection of software packages and specifications created by Sun Microsystems in the mid-1990s. The Java platform and language began as an internal project at Sun in 1990, and the goal was to develop a platform independent language that could be used for programming embedded software for consumer devices such as toasters and microwave ovens. A small project team was formed, consisting of James Gosling, Patrick Naughton, Mike Sheridan and others. Their mission was to create a cross-platform language that would produce code that would run on a variety of CPUs in different environments and could be easily ported to all types of devices.

They originally considered using the C++ programming language, but rejected it, as they believed that it used too much memory in the development of embedded systems, and that its complexity led to developer errors. Further, they had concerns about the garbage collection in C++ as well as the portability of the language.

The invention of the World Wide Web in the early 1990s had a major impact on the development of Java. Portability was an a key requirement for the Web, and it was obvious to Gosling and the others that the problem of portability encountered when creating software for embedded controllers are also encountered when creating software for the Internet. The latter is a distributed system consisting of many types of computers, operating systems and central processing units. This led to a change in the focus of Java from embedded software to Internet programming.

Gosling initially tried to modify and extend C++ but abandoned it in favour of a new language. The new language was initially called *Oak*, but it was renamed to "Java" in 1995. The syntax of the language is adapted from the C programming language, and the object-oriented features are adapted from C++. The team demonstrated part of the new platform in mid-1992 including the operating system, the Oak language, the libraries and the hardware. They piloted the new platform in the development of a personal digital assistant (PDA) later that year, and in mid-1994 the platform was targeted towards the World Wide Web. The word "Oak" is a trademark of Oak Technology, and so the language was renamed to Java. The first public release of the language was at the SunWorld conference in mid-1995. Sun established the JavaSoft group to continue the development of the technology.

Java has had a major impact on the Internet and has simplified web programming. It introduced a fundamentally new network program called the *Java applet*, and this is a special type of Java program that is designed to be transferred over the Internet and automatically executed by a web browser. An applet is downloaded on demand without further user interaction. In other words, if a user clicks a link that contains a Java applet, then the applet will be automatically downloaded and run in the browser. *An applet is a dynamic self-executing program that is initiated by the server and runs on the client machine.* The creation of the applet fundamentally changed Internet programming as it brought the world of dynamic self-executing programs into cyberspace. Java includes security features to prevent the applet from acting as malicious software by restricting the applet to the Java execution environment and preventing it from accessing other parts of the computer. This allows applets to be downloaded safely with confidence that no harm will be done and that security will not be breached.

Portability is a key concern for the Internet given the diverse range of computers and operating systems connected to it. The same code must work on all computers, as it is not practical to have different versions of the applet for all possible computers. This requires a way to generate portable executable code, and Java solves the portability problem in that the output of a Java compiler is *bytecode* and not executable code. Bytecode is an optimized set of instructions designed to be run by the *Java Virtual Machine* (JVM). The JVM is the Java run time system, and the original JVM was designed as an interpreter for bytecode.

The translation of Java into bytecode solves the portability problems for web-based programs and allows the program to be executed in many different environments. The JVM needs to be implemented for each environment, and once this is done, any Java program may run on it. The JVM will vary from platform to platform, but it understands bytecodes and executes the program in its run time environment. The fact that the program is run by the JVM helps to make it secure, since the JVM is in control, and it can prevent the program from generating side effects outside the system.

An interpreted program is generally slower than compiled executable code. However, for Java the difference is not as large due to the fact that the bytecode has been highly optimized. Further, a *just in time* (JIT) compiler may be part of the JVM, and it can compile selected portions of the bytecode into executable code in real time. A JIT compiler compiles code as it is needed during execution.

There have been many changes to Java since the Java Development Kit (JDK) was released in early 1996. There have been many additions of classes and packages to the standard library, and the evolution of Java is controlled through the Java Community Process (JCP) which allows users to propose additions and changes to the Java platform. The Java run time environment is on millions of personal computers, and Java ME (Java Platform Micro Edition) has become popular in mobile devices. Google's Android operating system uses the Java language.

28.2   Java Programming Language

Java is a general-purpose object-oriented programming language designed by
Gosling and others at Sun Microsystems. Its syntax was influenced by the C and
C++ programming language. The following is an example of the Hello World
program written in Java:

```
class HelloWorld
{
  public static void main (String args[])
  {
    System.out.println (``Hello World!'');
  }
}
```

The reader is referred to [AGH:13] for detailed information on the Java programming language.

28.3   Java Virtual Machine

Java was designed with portability in mind, and the objective is to create a cross-
platform computing environment that allows a program to be written once and
executed anywhere. Platform independence is achieved by compiling the Java code
into Java bytecode, where the bytecode is an optimized set of instructions.

The bytecode is then run on a Java Virtual Machine (JVM) that interprets and
executes the bytecode. The JVM is specific to the native code on the host hardware.
The problem with interpreting bytecode is that it is slow compared to traditional
compilation. However, Java addresses this problem with *just in time* compilation.
Java also provides automatic garbage collection which protects programmers who
forget to deallocate memory (thereby causing memory leaks).

A *native compiler* compiles the source code into the equivalent executable code.
Other compilers may compile the source code to the object code of a virtual
machine, and the translator module of the virtual machine translates each bytecode
of the virtual machine to the corresponding native machine instruction. That is,
the virtual machine translates each generalized machine instruction into a specific
machine instruction (or instructions) that may then be executed by the processor on
the target computer.

Most computer languages such as C require a separate compiler for each
computer platform (i.e. computer and operating system). However, a language such
as Java comes with a virtual machine for each platform. This allows the source code
statements in these programs to be compiled just once, and they will then run on
any platform.

Chapter 29
C.A.R. Hoare

Charles Anthony Richard (C.A.R. or Tony) Hoare is a British computer scientist who has made fundamental contributions to computing including the quicksort algorithm, the axiomatic approach to program semantics and programming constructs for concurrency (Fig. 29.1).

He was born in 1934 and studied philosophy and the classics at Merton College, at Oxford University in England. He obtained his undergraduate degree in 1956 and then pursued a graduate course in statistics at Oxford. He studied Russian at the Royal Navy during his National Service in the late 1950s, and he went to the Soviet Union in 1959 as a graduate student to study computer translation of human languages at the Moscow State University. He discovered the well-known sorting algorithm quicksort while investigating efficient ways to look up words in a dictionary.

He returned to England in 1960 and worked as a programmer for Elliot Brothers. This was a small company that manufactured scientific computers, and he led a team that implemented an early commercial compiler for the Algol 60 programming language.

He was promoted to the position of chief engineer and led a large project team to implement an operating system. This project was a disaster, but he managed to steer a recovery. He then took up the position of chief scientist at the computer research division of the company.

He became professor of computer science at Queens University in Belfast in 1968, and he became interested in techniques to assist with the implementation of operating systems. He was especially interested in to see if advances in programming theory and languages could assist with the problems of concurrency. He built up a teaching and research department and published material on the use of assertions to prove correctness of computer programs and the axiomatic semantics of programming languages.

G. O'Regan, *Giants of Computing: A Compendium of Select, Pivotal Pioneers*,
DOI 10.1007/978-1-4471-5340-5_29, © Springer-Verlag London 2013

Fig. 29.1 C.A.R. Hoare

Hoare argues for simplicity in language design and argued against languages such as Algol 68 and Ada, which attempted to be all things to all people. His comments on the flaws in the design of the Algol 68 language are:

> Algol 60 was a great achievement in that it was a significant advance over most of its successors.

His views on the importance of simplicity in design are:

> There are two ways of constructing a software design. One way is to make it so simple that there are obviously no deficiencies, and the other way is to make it so complex that there are no obvious deficiencies. The first method is far more difficult.

Hoare built upon Robert Floyd's work on applying assertions to flow charts, and he showed how the semantics of a small programming language could be defined with precondition and postcondition assertions. He showed how this could be applied to prove partial correctness of programs.

He moved to Oxford University in 1977 following the death of Christopher Strachey (who was well known for his work in denotational semantics). He built up the programming research group at Oxford, and this internationally respected group produced the Z formal specification language and Communicating Sequential Processes (CSP). A Z specification is a formal mathematical description of the requirements of the proposed system, whereas CSP is a mathematical approach to the study of communication and concurrency. CSP is used to specify and design computer systems that continuously interact with their environment.

He has received numerous awards in recognition of his contributions to the computing field. He received the Turing Award in 1980 for fundamental contributions to the definition and design of programming languages. He was elected a fellow of the Royal Society in 1982 and received a knighthood in 2000 for his service to computer science. He received the IEEE John von Neumann medal in 2011. He became a fellow of the Computer History Museum in California 2006. He is currently a principal researcher at Microsoft Research in Cambridge in England and is an emeritus professor at Oxford University.

29.1 Hoare Logic

Hoare logic is a formal system of logic used for mathematical reasoning about the correctness of computer programs. It includes axioms for the constructs of a simple programming language, and it is employed in programming language semantics and in program verification. It was originally published in Hoare's 1969 paper "An axiomatic basis for computer programming" [Hor:69].

The central feature of Hoare logic is the *Hoare triple*, and this describes how the execution of a fragment of code changes the state. It is of the form

$$P\{Q\}R$$

where P and R are assertions and Q is a program or command. The predicate P is called the *precondition*, and the predicate R is called the *postcondition*.

Hoare realized that Floyd's approach[1] of using assertions in flow charts provided an effective method for proving the correctness of programs. He built upon Floyd's work to include all of the familiar constructs of high-level programming languages. This led to the axiomatic approach to defining the semantics of every statement in the language with axioms and proof rules.

Definition 29.1 (Partial Correctness) The meaning of the Hoare triple P{Q}R is that whenever the predicate P holds of the state before the execution of the command or program Q, then the predicate R will hold after the execution of Q. The brackets indicate partial correctness as if Q does not terminate then R can be any predicate.

Total correctness requires Q to terminate, and at termination, R is true. Termination needs to be proved separately. Hoare (and Dijkstra) believed that the starting point of a project should always be the specification, and *that the proof of the correctness of the program should be developed hand in hand with the program itself.* That is, one starts off with a mathematical specification of what a program is supposed to do, and mathematical transformations are applied to the specification until it is a program that can be executed. The resulting program is correct by construction.

29.2 Axiomatic Definition of Programming Languages

An assertion is a property of the program's objects: e.g. the assertion $(x - y > 5)$ is a predicate that may or may not be satisfied by a state of the program during execution. For example, it is true in the state in which the values of the variables x and y are 7 and 1, respectively, and false in the state in which x and y have values 4 and 2, respectively.

---

[1]Robert Floyd was discussed in an earlier chapter.

The first article on program verification using assertions was by Robert Floyd as discussed in an earlier chapter. The paper was concerned with assigning meaning to programs, and the program was expressed by flow charts, and an assertion was attached to the flow chart. The assertion would be true during program execution, whenever execution reached that edge of the flow chart. Hoare refined and built upon Floyd's work, and he proposed a logical system for proving properties of program fragments. The Hoare triple was introduced with well-formed formulae of the form:

$$P \{Q\} R$$

where P is the precondition, Q is the program fragment, and R is the postcondition. The precondition P is a predicate (input assertion), and the postcondition R is a predicate (output assertion). The braces separate the assertions from the program fragment. The well-formed formula $P\{Q\}R$ is itself a predicate, i.e. either true or false. This notation expresses the partial correctness of Q with respect to P and R.

Definition (Partial Correctness) A program fragment Q is partially correct for precondition P and postcondition R if and only if whenever Q is executed in any state in which P is satisfied and the execution terminates, then the resulting state satisfies R.

It is required to prove that the postcondition R is satisfied if the program terminates. Partial correctness is a useless property unless termination is proved, as any nonterminating program is partially correct with respect to any postcondition.

Definition (Total Correctness) A program fragment Q is totally correct for precondition P and postcondition R if and only if whenever Q is executed in any state in which P is satisfied the execution terminates and the resulting state satisfies R.

The proof of total correctness requires proof that the postcondition R is satisfied and that the program terminates. Total correctness is expressed by $\{P\}Q\{R\}$. Dijkstra's calculus of weakest preconditions is based on total correctness, whereas Hoare's approach is based on partial correctness.

Hoare's axiomatic theory of programming languages consists of axioms and rules of inference to derive certain pre-post formulae. The meaning of several constructs of a simple programming language is presented here in terms of pre-post semantics (Table 29.1).

29.3   Communicating Sequential Processes

The objectives of the process calculi [Hor:85] are to provide mathematical models which provide insight into the diverse issues involved in the specification, design and implementation of computer systems which continuously act and interact with their environment. These systems may be decomposed into subsystems, which interact with each other and their environment. The basic building block is the *process*,

Table 29.1 Axioms of programming language constructs

Statement	Meaning
Skip $P \{skip\} P$	Skip does nothing, and this instruction guarantees that whatever condition is true on entry to the command is true on exit from the command
Assignment $P_e^x \{x := e\} P$	The meaning of this statement is that P will be true after execution if and only if the predicate P_e^x with the value of x replaced by e in P is true before execution (since x will contain the value of e after execution)
Compound $\dfrac{P\{S_1\}Q,\; Q\{S_2\}R}{P\{S_1;\; S_2\}R}$	The execution of the **compound** statement involves the execution of S_1 followed by S_2. Its correctness with respect to P and R is established by proving that the correctness of S_1 with respect to P and Q and the correctness of S_2 with respect to Q and R
Conditional $\dfrac{P\wedge B\{S_1\}Q,\; P\wedge\neg B\{S_2\}Q}{P\{\text{if } B \text{ then } S_1 \text{ else } S_2\}Q}$	The execution of the **conditional** statement involves the execution of S_1 or S_2. The execution of S_1 takes place only when B is true, and the execution of S_2 takes place only when $\neg B$ is true
	Its correctness with respect to P and Q is established by proving that both S_1 and S_2 are correct with respect to P and Q
	S_1 is executed only when B is true, and S_2 is executed only when B is false. Thus it is required to prove the correctness of S_1 with respect to $P\wedge B$ and Q, and the correctness of S_2 with respect to $P\wedge\neg B$ and Q
While loop $\dfrac{P\wedge B\{S\}P}{P\{\text{while } B \text{ do } S\}P\wedge\neg B}$	The property P is termed the loop invariant as it remains true throughout the execution of the loop
	The execution of the **while loop** is such that if the truth of P is maintained by one execution of S, then it is maintained by any number of executions of S. The execution of S takes place only when B is true, and upon termination of the loop $P \wedge \neg B$ is true
	Loops may fail to terminate, and there is therefore a need to prove termination

which is a mathematical abstraction of the interactions between a system and its environment.

Processes may be assembled into systems, execute concurrently or communicate with each other. Process communication may be synchronized, with one process outputting a message simultaneously to another process inputting a message. A process may be specified recursively, and resources may be shared among several processes. CSP enriches the understanding of communication and concurrency and obeys nice mathematical laws.

The expression $(a \rightarrow P)$ in CSP describes a process which first engages in event a and then behaves as process P. For example, a vending machine that serves one customer before breaking may be defined as:

$$(\text{coin} \rightarrow (\text{choc} \rightarrow \text{STOP}))$$

A recursive definition is written as $(\mu X):A \bullet F(X)$, where A is the alphabet of the process. The behaviour of a simple chocolate vending machine is given by the following recursive definition:

$$VMS = \mu X : \{coin, choc\} . (coin \rightarrow (choc \rightarrow X))$$

The simple vending machine has an alphabet of two symbols, namely, coin and choc, and the behaviour of the machine is such that when a coin is entered into the machine, a chocolate is then provided. This machine repeatedly serves chocolate in response to a coin.

The definition of the intended behaviour of the process is termed its specification, and the implementation of a process can be proved to meet its specification (i.e. *P* **sat** *S*). Processes can be assembled together to form systems, and the processes interact with each other and their environment. The environment may also be described as a process, and the complete system is regarded as a process with its behaviour is defined in terms of the component processes. CSP is described in more detail in [ORg:06, Hor:85].

Chapter 30
Herman Hollerith

Herman Hollerith was an American statistician, inventor and entrepreneur. He founded a tabulating machine company in 1896, and this company later merged with another company to become the Computing Tabulating Recording Company. This company was renamed to International Business Machines (IBM) in 1911 (Fig. 30.1).

He was born in Buffalo, New York, in 1860 and was the son of German immigrants. He attended the Columbia University School of Mines and qualified as a mining engineer in 1879. His undergraduate record was excellent, and after graduation he became an assistant to one of his teachers, Dr. William Trowbridge, at Columbia University.

Trowbridge was appointed as a consultant to the US Census Bureau, and Hollerith joined the Census Bureau as a statistician. His goal was to assist the Census Bureau in processing the large amount of data generated from the 1880 population census[1] of the United States. This stimulated his interest in automating the tabulation of the census data and in manipulating the data mechanically. Dr. John Billings[2] believed that there must be a mechanical way to do this along the lines of the Jacquard Loom, and he discussed his idea with Hollerith. Billings believed that the holes in the card should be able to tabulate the population and similar statistics in a manner similar to that done by a card in the Jacquard loom that regulates the pattern to be woven in the fabric.

Hollerith took a position at the Massachusetts Institute of Technology in 1882 and taught mechanical engineering. He remained interested in the problem of automating the tabulated census data and examined the workings of the Jacquard loom to see if it could assist. The main feature of the Jacquard loom that he

---

[1] The processing of the 1880 population census was done manually and took several years to complete.

[2] Dr. John Billings did pioneering work on statistics for the 1880 and 1890 population census. He was also a distinguished surgeon. He proposed the idea of mechanization of census data tabulation to Hollerith.

G. O'Regan, *Giants of Computing: A Compendium of Select, Pivotal Pioneers*,
DOI 10.1007/978-1-4471-5340-5_30, © Springer-Verlag London 2013

Fig. 30.1 Herman Hollerith
(Courtesy of IBM archives)

considered applicable was its use of punched cards, as these were an efficient way of storing information. He also observed that it should be possible to punch information onto a card, as a conductor might do on a train. Hollerith conducted a number of experiments at the institute and initially employed a paper tape rather than cards. A pin could go through a hole in the tape and complete an electrical circuit, and Hollerith later replaced paper with cards as these offered a more effective solution.

He took a position at the US Patent Office in 1884, and he patented his invention later that year. He understood the importance and potential value of protecting his inventions, and he was granted a total of 30 patents during his career. He did further work on methods to convert the information on punched cards into electrical impulses, which could activate mechanical counters. He initially employed the ticket punch used by conductors on the railway to punch holes but later designed a more effective punch for his system.

Hollerith formed the Tabulating Machine Company in Washington, D.C., in 1896, and this was the first electric tabulating machine company. His system was first tested on mortality systems in Baltimore in 1887. It was predicted that the 1890 population census would take several years to process, and the US Census Bureau recognized that its existing methodology was no longer fit for purpose. It held a contest to find a more efficient and cost effective system, and Hollerith's system was the clear winner.

His punched card Tabulating Machine used an electric current to sense holes in punched cards, and it kept a running total of the data. This allowed the statistics to be recorded by electrically reading and sorting punched cards (Fig. 30.2). The new methodology enabled the results of the 1890 census to be available in a couple of months rather than years and led to a saving of millions of dollars in tabulating the census data. The population was recorded to be over 62 million in 1890. Hollerith's system was later used to tabulate the census data in several other countries including Russia and Canada.

Fig. 30.2 Hollerith's tabulator (1890) (Courtesy of IBM archives)

Hollerith's Tabulating Machine Company later merged with the International Time Recording Company to form the Computing Tabulating Recording Company (CTR) in 1911. Thomas Watson joined the company in 1914 when the company was going through difficulties. He turned the company around, and the company changed its name to *International Business Machines* (*IBM*) in 1924. IBM has been in business for over 100 years and remains highly successful.

Hollerith published the details of his tabulating invention and submitted it for a doctorate degree at Columbia University. His Ph.D. was awarded in 1890, and he received several awards in recognition of this excellent invention. These included the Elliot Cresson Medal from the Franklin Institute of Philadelphia in 1890.

Hollerith's punched cards and tabulating machine was a step forward towards automated computation. His device could automatically read information that had been punched on to the card, but the machine was limited to tabulation and could not be used for more complex computations.

The tabulator was an important precursor to the modern computer. Punched card technology remained in use on computers up to the late 1970s.

Chapter 31
Watts Humphrey

Watts Humphrey was an American software engineer and vice president of technical development at IBM. He made important contributions to the software engineering field and was known as the *father of software quality*. He dedicated much of his career to addressing the problems of software development including schedule delays, cost overruns, software quality and productivity (Fig. 31.1).

He was born in Michigan in 1927 and faced challenges in learning as a child as he had dyslexia. He served in the US Navy and completed a bachelor's degree in physics at the University of Chicago in 1949. He obtained a Master's degree in physics from the Illinois Institute of Technology (IIT) and an MBA from the University of Chicago.

He took a position with Sylvania in Boston in the early 1950s, and he became manager of the circuit design group in the company. He recognized the importance of planning and management early in his career, and he later made important contributions to the management aspects of software development at IBM and the Software Engineering Institute (SEI). He joined IBM in 1959 initially as a hardware architect, but most of his IBM career was in management. He was eventually to become a vice president of technical development, where he oversaw 4,000 engineers in 15 development centres in over 7 countries. He was influenced by others at IBM including Fred Brooks who was a project manager of the IBM/360 project, Michael Fagan who developed the Fagan Inspection Methodology and Harlan Mills who developed the Cleanroom methodology. Humphrey ran the software quality and process group at IBM towards the end of his IBM career and became very interested in software quality.

He retired from IBM in 1986 and joined the newly formed SEI at Carnegie Mellon University. He made a commitment to change the software engineering world by developing sound management principles for the software industry. The SEI has largely fulfilled this commitment, and it has played an important role in enhancing the capability of software organizations throughout the world.

The SEI had a contract from the Department of Defence (DOD) to provide guidance to the military in the selection of capable software subcontractors. This evolved into the book *Managing the Software Process* [Hum:89] which describes

G. O'Regan, *Giants of Computing: A Compendium of Select, Pivotal Pioneers*,   143
DOI 10.1007/978-1-4471-5340-5_31, © Springer-Verlag London 2013

Fig. 31.1 Watts Humphrey
(Courtesy of Watts
Humphrey)

technical and managerial topics essential for good software engineering. The book was influenced by the ideas of Deming and Juran in statistical process control.

Humphrey established the software process program at the SEI, and this led to the development of the software Capability Maturity Model (CMM) and its successors. The CMM is a framework to help an organization to understand its current process maturity and to prioritize improvements. He introduced software process assessment and software capability evaluation methods, and these include CBA/IPI and CBA/SCE. The CMM model and the associated assessment methods were widely adopted by organizations around the world, and their successors are the CMMI Model and SCAMPI appraisal methodology.

Humphrey focused his later efforts to developing the Personal Software Process (PSP) and the Team Software Process (TSP). These are approaches that teach engineers the skills they need to make and track plans and to produce high-quality software with zero defects. The PSP helps the individual engineer to collect relevant data for statistical process control, whereas the TSP focuses on teams, and the goal is to assist teams to understand and improve their current productivity and quality of their work.

He has received many awards for his contributions to the computing field. He was named the first SEI fellow in 1995 in recognition of his outstanding contribution to the software quality field. He received the 2003 National Medal in Technology from President George Bush and was named an ACM fellow in 2009 for his outstanding contributions to computing and information technology. He is the author of 12 books in the software engineering field. He died in 2010.

31.1 Software Process Improvement

Software process improvement is concerned with practical action to improve the processes in the organization, to ensure that they meet business goals more effectively. For example, the business goal might be to improve process performance to enable high-quality software products to be developed and delivered faster to the market place. The origins of the software process improvement field go back to the manufacturing sector and to Walter Shewhart's work on statistical process control in the 1930s.

His work was later refined by Deming and Juran, who argued that high-quality processes are essential to the delivery of a high-quality product. They argued that the quality of the end product is largely determined by the processes used to produce and support it. Therefore, there is a need to focus on the process as well as the product itself, and high-quality products need to be built from processes that are fit for purpose. Their approach transformed struggling manufacturing companies with quality problem to companies that could consistently produce high-quality products. This led to cost reductions and higher productivity, as less time was spent in reworking defective products. Their focus was on the manufacturing process and in reducing its variability [ORg:02].

This work was later applied to the software quality field by Watts Humphrey and others at the SEI, leading to the birth of the software process improvement field. Humphrey asked questions such as:

– How good is the current software process?
– What must I do to improve it?
– Where do I start?

Software process improvement initiatives support the organization in achieving its key business goals such as delivering software faster to the market, improving quality and reducing or eliminating waste. The objective is to work smarter and to build software better, faster and cheaper than competitors. Software process improvement makes business sense and provides a tangible return on investment.

Humphrey recognized that a software process improvement initiative involves change to the way that work is done and therefore needs top management support to be successful. It requires the involvement of the software engineering staff, and changes to the process are made based on an understanding of the strengths and weaknesses of the process. Every task and activity can be improved and so change is continuous. The processes need to be reinforced with training, and audits need to be conducted to ensure process fidelity. Software quality is improved by improving the software processes.

The Software Engineering Institute (SEI) developed the Capability Maturity Model (CMM) in the early 1990s as a framework to help software organizations to improve their software process maturity and to implement best practice in software and systems engineering. The SEI believes that there is a close relationship between the maturity of software processes and the quality of the delivered software product. The first version of the CMM was released in 1991, and its successor is the Capability Maturity Model Integration (CMMI®) [CKS:11].

The SEI maintains data on the benefits that organizations have achieved from using the CMM and CMMI. It has measured the improvements in several categories such as cost, schedule, productivity, quality, customer satisfaction and a return on investment. The results in Table 31.1 are from 25 organizations [SEI:06].

Table 31.1 Benefits of software process improvement (CMMI)

Improvements	Median	#Data points	Low	High
Cost	20 %	21	3 %	87 %
Schedule	37 %	19	2 %	90 %
Productivity	62 %	17	9 %	255 %
Quality	50 %	20	7 %	132 %
Customer satisfaction	14 %	6	−4 %	55 %
ROI	4.7:1	16	2:1	27:1

For example, *Northrop Grumman Defense Systems* met every milestone (25 in a row) with high quality and customer satisfaction; *Lockheed Martin* reported an 80 % increase in software productivity over a 5-year period when it achieved CMM level 5. *Siemens* (*India*) reported an improved defect removal rate from over 50 % before testing to over 70 % before testing and a post-release defect rate of 0.35 defects per KLOC. *Accenture* reported a 5:1 return on investment from software process improvement activities.

Software process improvement provides a return on investment and makes business sense. More detailed information is in [ORg:10].

31.2   Capability Maturity Model Integrated (CMMI)

The SEI developed the CMM in the early 1990s as a framework to help software organizations improve their software process maturity. The CMMI is its successor and appeared early in the new millennium. The CMM and CMMI are used to implement best practice in software and systems engineering. The SEI and other quality experts believe that there is a close relationship between the maturity of software processes and the quality of the delivered software product.

Software companies need to have high-quality software processes to develop high-quality software, and the CMMI is a framework to assist an organization in the implementation of best practice in software and systems engineering. This internationally recognized model for software process improvement is used in thousands of organizations around the world. It enables management to identify which processes to improve, and how to improve them.

The process is an abstraction of the way in which work is done and is seen as the glue (Fig. 31.2) that ties people, procedures and tools together. A *process* is a set of practices or tasks performed to achieve a given purpose. It may include tools, methods, material and people. An organization will typically have many processes in place for doing its work, and the object of process improvement is to improve the key processes to meet business goals more effectively.

The CMMI consists of five maturity levels with each maturity level (except level one) consisting of several process areas. Each process area consists of a set of goals, which must be implemented for the process area to be satisfied. The goals

Fig. 31.2 Process as glue for
people, procedures and tools

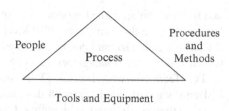

are implemented by practices related to that process area, and processes need to be defined and documented. The users of the process need to receive appropriate training to enable them to carry out the process, and processes need to be enforced by independent audits.

The emphasis on level two of the CMMI is on maturing management practices such as project management, requirements management and configuration management The emphasis on level three of the CMMI is to mature engineering and organization practices. This maturity level includes peer reviews and testing, requirements development and software design. Level four is concerned with ensuring that key processes are performing within strict quantitative limits and adjusting processes, where necessary, to ensure that performance is within these limits. Level five is concerned with continuous process improvement that is quantitatively verified.

Maturity levels may not be skipped in the staged implementation of the CMMI. There is also a continuous representation of the CMMI that allows the organization to focus on improvements to key selected processes. However, in practice, it is often necessary to implement several of the level two process areas before serious work can be done on implementing a process at a higher maturity level. The use of metrics [Fen:95, Glb:76] becomes more important as an organization matures, as metrics allow the performance of an organization to be objectively judged.

The CMMI helps companies deliver high-quality software systems that are consistently on time and consistently meet business requirements. There is more detailed information on software process improvement and the CMMI in [ORg:10].

31.3   PSP and TSP

The Personal Software Process (PSP) is a disciplined data-driven software development process that is designed to help software engineers understand and to improve their personal software process performance. It was developed by Watts Humphrey at the SEI, and it helps engineers to improve their estimation and planning skills and to reduce the number of defects in their work. This enables them to make commitments that they can keep and to manage the quality of their projects.

The PSP is focused on the work of the individual engineer, and it includes methods and tools that enable the engineer to produce a product on time, on budget and with the right quality. PSP helps the engineer to get the required data and to focus on statistical process control. The process has three levels with each level offering a different focus. PSP level 1 is focused on estimation and planning (including time, size and defect estimates); PSP level 2 is focused on design, code reviews and quality management; and PSP level 3 is focused on larger projects.

The Team Software Process (TSP) was developed by Watts Humphrey at the SEI and is a structured approach designed to help software teams understand and improve their quality and productivity. Its focus is on building an effective software development team, and it involves establishing team goals, assigning team roles as well as other teamwork activities. Team members must already be familiar with the PSP.

The teams are *self-directed* and define their own work processes and produce their own plan. They set their own quality goals and build plans to achieve these goals. They track and report their work (including hours worked, defects found per phase, inspection examination rates and defect densities). The teams share team management responsibilities. TSP enables teams to adopt a data-driven approach to software development.

Chapter 32
Kenneth Iverson

Kenneth Iverson was a Canadian computer scientist who developed the APL programming language, and he made important contributions to mathematical notation and programming language theory (Fig. 32.1).

He was born in Alberta in 1920 and showed an early aptitude for mathematics. He left school at an early age to work on his parents' farm, and he served as a flight engineer in the Royal Canadian Air Force during the Second World War. He completed his high school diploma by taking correspondence courses organized by the Canadian Legion.

He enrolled in Queens University in Ontario and studied mathematics and physics. He graduated with a bachelors' degree in 1950 and obtained a Masters' degree in mathematics from Harvard University in 1951. He began working at Harvard with Howard Aiken (who designed the Harvard Mark I computer) and with Wassily Leontief (a Russian American economist who did research on input–output analysis and on how changes in one economic sector can have an effect on other economic sectors). Iverson moved to the Department of Engineering and Applied Physics at Harvard, and his Ph.D. was in applied mathematics and extended Leontief's work. The experimental work for his Ph.D. degree involved writing a software package for matrix routines on the Harvard Mark I computer. His Ph.D. was awarded in 1954.

He taught at Harvard for several years and became interested in a precise mathematical notation for the expression of algorithms, as he believed that existing techniques were inadequate. He began to develop his own notation (informally called "Iverson's notation" but later renamed to APL). The notation was based on arrays and operators.

He joined IBM's Research Center at Yorktown Heights, New York, in 1960 and published the classic book *A Programming Language* in 1962 [Ive:62]. This book gave the name "APL" to the notation that he had developed for algorithms, and it later became a programming language. He extended the language at IBM to the extent that it could be used for the description of algorithms and systems.

APL was used for the formal description of the IBM System/360 series of machine architecture and functionality, which was written in 1963. The first

Fig. 32.1 Ken Iverson

implementation of APL was in FORTRAN in 1965 on an IBM 7090 computer. Iverson continued to work on APL throughout his career at IBM.

He left IBM in 1980 and returned to Canada to work for I.P. Sharp Associates (a Canadian consulting company). He retired from the company in 1987 and worked on a more modern version of APL. The goal was to make improvements to APL and to include more advanced features such as functional programming, arrays of variables and support for multiple instruction and multiple data (MIMD) parallel operations. The first implementation of this language was called J, and it was released in 1990.

He received various awards in recognition of his contributions. He became an IBM fellow in 1970, and he received the ACM Turing Award in 1979. He died of a stroke in 2004.

32.1   Turing Award (Notation as a Tool of Thought)

Iverson strongly believed in the power of *notation as a tool of thought*, and this was the title of his 1979 Turing Award lecture. He quoted Whitehead's influential statement on the characteristics of a good notation [Whi:11] in the introduction to his Turing Award lecture.

> By relieving the brain of all unnecessary work, a good notation sets it free to concentrate on more advanced problems, and in effect increase the mental power of the race.

Iverson states in this paper that while mathematical notation is the best known and best developed language as a tool of thought that it has deficiencies. He argues that mathematical notation lacks universality, and that it must be interpreted differently according to the context. Programming languages were designed for the purpose of directing machines and are executable and unambiguous. However, in other respects they are inferior to mathematical notation as a tool of thought.

Iverson's thesis is that the advantages of executability and universality found in programming languages can be effectively combined with mathematical notation to form a single coherent language. He outlines several desirable characteristics of a good notation such as:

- Ease of expressing constructs that arise in a problem domain
- Unambiguous
- Terse and economical
- Amenable to formal proof

He argues that APL, i.e. the executable language that he has developed, is a powerful and concise programming language that combines the richness of mathematical notation with the features of a programming language. This leads to our discussion of APL.

32.2 APL

APL is a powerful, expressive and concise programming language developed by Kenneth Iverson. It is an *array-oriented language* and is based on the mathematical notation for manipulating arrays developed by Iverson. The language is used for scientific and financial applications and influenced the development of spreadsheets, functional programming and computer maths packages. Its features include:

- A concise language that uses symbols instead of words
- Applies functions to entire arrays without using loops
- Solution focused
- Emphasizes expression of algorithm independent of machine architecture
- Facilitates problem solving at a high level of abstraction

APL distinguishes between functions and operators. A *function* takes arrays (containing variables, constants or expressions) as arguments and returns arrays as results. An *operator* is similar to a higher-order function that takes a function as an argument; it takes a functions or arrays as arguments and returns functions as a result.

The APL environment is called a *workspace*, and the user can define programs and data in the workspace. In other words, the data values also exist outside the programs, and the user can manipulate the data without defining a program. The user can save the workspace with all values and programs. APL uses a set of non-ASCII symbols, and its terse notation allows algorithms to be concisely expressed.

The basic data structure in APL is the vector, and APL programs rarely use loops or conditional execution, as everything is transformed into vector/matrix operations. For example, the vector values 1 3 5 7 are assigned to N by the statement:

$$N \leftarrow 1\ 3\ 5\ 7$$

Similarly, the statement that adds 3 to all values in N giving (4 6 8 10) and then prints them is given by

$$N + 3$$

The statement that computes the sum of the values of N, and then prints it (i.e. $1 + 3 + 5 + 7 = 16$), is given by

$$+/N$$

APL programs are generally much shorter than the equivalent program in other programming languages. For example, consider the sum of all integers from 1 to 100. This would require several lines in a high-level programming language such as C or Pascal, whereas in APL it is given by the one line program:

$$+/i100$$

The meaning of this statement is that the program first generates a vector containing a 100 successive integer elements ranging from 1 to 100, and the program then computes the sum of these vector elements. The first part is done by the statement $i100$ where i is an operator that takes an integer argument and returns a vector containing 100 successive integer elements ranging from 1 to 100. The second part involves computing the sum by using the reduction operator "/" which has an argument on its left and right side. The reduction operator places the operator on its left-hand side with every two successive elements of the vector on its right-hand side. In other words,

$$+/i100 = 1 + 2 + 3 + \cdots + 99 + 100$$

Chapter 33
Ivar Jacobson

Ivar Jacobson is a Swedish computer scientist who is regarded as one of the fathers of components and component architecture. He is well-known for his work on SDL; use cases, use case driven development; the unified modeling language (UML); the Rational Unified Process (RUP); and aspect-oriented software development (Fig. 33.1).

He was born in Ystad, Sweden, in 1939 and obtained his Master's degree in electrical engineering from the Chalmers Institute of Technology in Gothenburg in 1962. He earned his Ph.D. from the Royal Institute of Technology in Stockholm in 1985.

He joined Ericsson shortly after completing his university education, and in 1967 he proposed the use of software components for the new generation of software-controlled telephone switches. He invented sequence diagrams and collaboration diagrams around this time and employed state transition diagrams to describe the message flows between components. He also invented use cases, and these are used to specify functional requirements. These diagrams later became part of the UML specification and modelling language.

Jacobson was one of the original developers of the System and Description Language (SDL), which later became a standard in the telecoms industry.

He started his own company, Objective Systems (later renamed to Objectory AB), in 1987. The telecommunications giant, Ericsson, took a majority stake in the company in 1991, and the company was sold to Rational Software in 1995. This allowed Jacobson to work with Grady Booch and James Rumbaugh, which led to the development of UML and the Rational Unified Process (RUP). Rational was taken over by IBM in 2003, and Jacobson left the company the following year.

He is the founder and chairman of Ivar Jacobson International, a global consulting company that works with companies around the world in implementing effective development practices to improve the performance of teams.

G. O'Regan, *Giants of Computing: A Compendium of Select, Pivotal Pioneers*,
DOI 10.1007/978-1-4471-5340-5_33, © Springer-Verlag London 2013

Fig. 33.1 Ivar Jacobson

33.1 Specification and Description Language (SDL)

SDL is a specification language that is used for the specification and implementation of real time systems. Its origins are in the telecommunications domain, and it began with a CCITT (later called ITU) study on telecoms and software technology. The study commenced in 1968 and focused on ways of handling stored program control switching systems. SDL was one of the first standardized notations for the specification of software systems, and it allows the system to be specified as a set of interconnected blocks.

It involved participation from several companies including GEC Telecommunications and Ericsson. Ivar Jacobson, Göran Hemdal and others at Ericsson shared their experience of the development of the AXE system at Ericsson, and this was to have an important influence on the development of the language. The software for AXE was organized by blocks with well-defined interfaces, and components and component-based development was used. The idea for software components was due to Jacobson.

The study recognized the need for a language for the specification of systems, and the first version of SDL was created in 1976. An object-oriented version was introduced in 1992, and SDL-2010 is the latest version of the language.

The language provides an equivalent graphical and textual representation of the system, and the graphical model of the system is usually presented. SDL covers structure, communication, behaviour, data and inheritance.

There are several support tools available for the language. These include commercial tools such as the Tau SDL suite from Telelogic and public domain tools such as JADE. The tools include features such as graphical editors, syntax checking and the code generation of C++ from the specification. The tools allow simulation of the specification. SDL is the telecoms standard (ITU-T Recommendations Z.100-Z.106).

33.2   Object-Oriented Software Engineering (OOSE)

OOSE was developed by Jacobson and others at Objectory AB in the early 1990s. It was the first object-oriented design methodology to employ *use cases* to drive the software design and development. The Objectory tool was used to implement OOSE.

The use case model describes the complete functionality of the system and serves as the central model of the system. It is the basis of the analysis, construction and testing phases. The objective of the analysis phase is to understand the system according to its functional requirements, and this involves identifying the objects and their interactions. The construction phase is concerned with design and implementation in source code, and components are used to implement objects. The testing phase is concerned with verifying the correctness of the system with respect to its specification.

OOSE is described in detail in [Jac:92]. Many of its concepts and notation have been incorporated into UML, and the OOSE notation and tools have been replaced by tools supporting UML and the Rational Unified Process.

33.3   Unified Modeling Language

The first object-oriented modelling language[1] emerged in the late 1960s, and by the early 1990s there was a plethora of object-oriented methods. The most popular methods included James Rumbaugh's *Object Modeling Technique* (OMT), the *Booch Method* developed by Grady Booch at Rational Software and Jacobson's *object-oriented software engineering* (OOSE) method. It was evident that there was a need for a single standardized unified modeling language with a formally defined syntax.

Rational Software hired James Rumbaugh from General Electric in 1994, and Rational took over Jacobson's company, Objectory AB, in 1995. Grady Booch was the chief scientist for Rational at this time, and this allowed Rumbaugh, Booch and Jacobson to work together to create a unified modeling language.

Their goal was not to create a new modelling language as such but to integrate the existing Booch Method, OMT and OOSE to form a single standardized modelling language. This involved simplifying or expanding the existing diagrams in several object-oriented methods. This included class diagrams, use case diagrams and activity diagrams. Their work led to a standardized unified language with a formal semantics of the language elements. Today, UML is an industry standard and the Object Management Group (OMG) is responsible for the evolution of the language.

---

[1]The first object-oriented programming language to emerge was Simul-67, which was developed in the 1960s at the Norwegian Computing Centre in Oslo by Ole Johan Dahl and Kristen Nygaard. This language introduced objects and classes.

Table 33.1 UML diagrams

Diagram	Description
Class	This shows the set of classes, interfaces and collaborations and their relationships. They address the static design view of the system
Object	This shows a set of objects and their relationships. They represent the static design view of the system but from the perspective of real cases
Use case	These describe the functional requirements from the user's point of view and describe a set of use cases and actors and the relationship between them
Sequence diagram	These diagrams show the interaction between a set of objects and messages exchanged between them. It emphasizes the time ordering of messages
Collaboration diagram	A collaboration diagram is a diagram that emphasizes the structural organization of objects that send and receive messages
Statechart	This shows a state machine consisting of states, transitions, events and activities. It addresses the dynamic view of a system and is important in modelling the behaviour of an interface or class
Activity diagram	This is a kind of statechart diagram that shows the flow from activity to activity of a system. It addresses the dynamic view of a system and is important in modelling the function and flow of control among objects
Component diagram	This diagram shows the organizations and dependencies among components. It addresses the static implementation view of a system
Deployment	This diagram shows the configuration of run time processing nodes and the components that live on them

UML is an expressive graphical modelling language for visualizing, specifying, constructing and documenting a software system. It provides several views of the software's architecture, which is essential in the development and deployment of systems. It has a clearly defined syntax and semantics for every building block of its graphical notation. Each stakeholder (e.g. project manager, developers and testers) has a different perspective and looks at the system in different ways at different times over the project's life. UML is a way to model the software system prior to implementation in some programming language, and the explicit visual model of the system facilitates communication among the various stakeholders. It has been employed in many domains including the banking sector, defence and telecommunications.

A UML specification involves building precise, complete and unambiguous models. Code may be generated from the models (using the available support tools) in a programming language such as Java or C++. The reverse is also possible, and thus it is possible to work in the graphical notation of UML or the textual notation of a programming language. UML expresses things that are best expressed graphically, whereas a programming language expresses things that are best expressed textually. The support tools are employed to keep both views consistent.

The nine key UML diagrams are described in Table 33.1. They are employed to provide a graphical visualization of the system from different viewpoints.

UML is described in more detail in [Jac:05]. It is often used as part of the Rational Unified Process (RUP).

33.4 Rational Unified Process

The origins of the *Rational Unified Process* (RUP) are in Objectory V1.0 (developed by Jacobson's company), Rumbaugh's OMT and the Booch method. The Objectory's processes were used to define the core processes in RUP, and an early version of RUP was released in 1998. RUP uses the visual modelling standard of UML, and a full description of the process is in [Jac:99]. These are the following:

- Use cases driven
- Architecture centric
- Iterative and incremental
- Employs a component-based architecture

It includes cycles, phases, workflows, risk mitigation, quality control, project management and configuration control. Software projects are often complex, and there are risks that requirements may be missed in the process, or that the interpretation of a requirement may differ between the customer and developer. Requirements are gathered as use cases, and the use cases describe the functional requirements from the point of view of the user of the system. The use case model describes what the system will do at a high level and provides a user focus in defining the scope of the project. Use cases drive the development process, and the developers create a series of design and implementation models that realize the use cases. The developers review each successive model for conformance to the use case model, and the testers verify that the implementation model correctly implements the use cases.

The architecture of the system is fundamental, and it includes the most significant static and dynamic aspects of the system. The architecture grows out of the use cases and from other pertinent factors such as the platform that the software is to be run on deployment considerations, legacy systems and nonfunctional requirements.

A commercial software project is often a large undertaking that may involve many person-years. It may take over 1 year to complete, and the work is decomposed into smaller slices or mini-projects, where each mini-project is a manageable chunk. Each mini-project is an iteration that results in an increment.

The unified process is a way to reduce risk in software development, as if the developers need to repeat the current iteration, then the organization loses only the misdirected effort of that iteration, rather than the entire project.

The waterfall software development has the disadvantage that the risk is greater towards the end of the project, where it is costly to undo mistakes from earlier phases. With an iterative process, the waterfall steps are applied iteratively. However, instead of developing the entire system in one step, an increment (i.e. a subset of the system functionality) is selected and developed, then another increment is

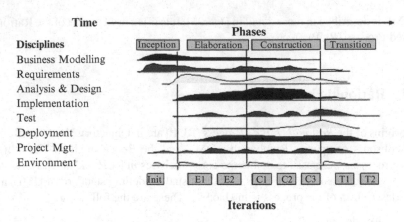

Fig. 33.2 Phases and workflows in the unified process

developed and so on. The earliest iterations address the areas with the greatest risk. Each iteration produces an executable release and includes integration and testing activities.

The Rational Unified Process consists of four phases. These are *inception*, *elaboration*, *construction* and *transition*. Each phase consists of one or more iterations, and each iteration consists of several workflows. The workflows may be requirements, analysis, design, implementation and test. Each phase terminates in a milestone with one or more project deliverables (Fig. 33.2).

The inception phase is concerned with the initial project planning and cost estimation and initial work on the architecture and functional requirements for the product. It also identifies and prioritizes the most important risks. The elaboration phase specifies most of the use cases in detail, and the system architecture is designed. The construction phase is concerned with building the product, the product contains all of the use cases agreed by management and the customer for the release. The transition phase covers the period during which the product moves into the customer site and includes activities such as training customer personnel, providing helpline assistance and correcting defects found after delivery.

Chapter 34
Steve Jobs

Steve Jobs was an American entrepreneur and inventor. He was the co-founder
(with Steve Wozniak) of Apple computers in 1976 and served as the chairman
and CEO of the company. He resigned from Apple in the mid-1980s following
disagreements at board level on the appropriate direction for the company, and he
founded NeXT, Inc., and later Pixar Animation Studios. He was the CEO of both
of these successful companies, and he returned to Apple (following its takeover
of NeXT) and transformed Apple to a highly successful and innovative company
(Fig. 34.1).

He was born in San Francisco, California, in 1955. His biological parents were
two University of Wisconsin graduate students who were unmarried,[1] and they gave
him up for adoption at birth. He was adopted by Paul and Clara Jobs and christened
Steven Paul Jobs. The family lived in Mountain View, California (adjacent to Silicon
Valley), and his mother was an accountant, and his father worked for an electronics
company. His father taught him elementary electronics in the family garage, and he
was soon able to take electronics apart and to correctly reconstruct it again.

He attended Homestead High School in Cupertino, California, and was introduced to Stephan Wozniak by a fellow student. Wozniak was attending the
University of California at Berkeley, and he subsequently joined Hewlett-Packard
to work on mainframe computers. Jobs spent a summer working at HP, and he
and Wozniak became friends. They shared a love of electronics and in designing
computers. Wozniak had designed several computers, and so he was the more
experienced of the two in computer design.

Jobs graduated from high school in 1972 and enrolled at Reed College in
Portland, Oregon. He dropped out after 6 months and spent the following 18 months

---

[1] His biological parents subsequently married and had a daughter (Mona Simpson) in 1957. Their
marriage later broke up and they divorced in 1962. Mona Simpson is a professor of English at the
University of California, LA, and is the author of several novels. Jobs met Mona for the first time
in 1985, and they developed a close friendship.

G. O'Regan, *Giants of Computing: A Compendium of Select, Pivotal Pioneers*,   159
DOI 10.1007/978-1-4471-5340-5_34, © Springer-Verlag London 2013

Fig. 34.1 Steve Jobs
at Macworld 2005

attending various creativity courses. Wozniak dropped out of Berkeley University after 1 year and joined Hewlett-Packard.

Jobs took a position with the video game company, Atari, in 1974. He took leave of absence to spend several months in India later that year, and he travelled in Uttar Pradesh and Himachal Pradesh seeking spiritual enlightenment from various Indian spiritual leaders. The period in India deeply influenced him, and he became a Zen Buddhist and experimented with psychedelic drugs such as LSD. On his return to the United States, he returned to Atari and worked on circuit board design.

Jobs and Wozniak began attending the Homebrew Computer Club in the mid-1970s, and they formed Apple Computers in 1976. The company began operations in Job's family garage, and their goal was to develop a user-friendly alternative to the existing mainframe computers produced by IBM and Digital. Wozniak was responsible for product development and Jobs for marketing.

The Apple I computer was introduced in 1977 with the Apple II following later that year. The Apple II was a major commercial success, and its sales were over $130 million. Apple became a public-quoted company in 1980 with a market value of over $1 billion. IBM introduced its own personal computer in 1981, and it became the dominant personal computer in the market. Apple introduced the technically superior Apple Macintosh in 1984, and although this machine was very successful, it was unable to break the IBM dominance of the market.

Jobs resigned from Apple in 1985 following serious disagreements at board level on the appropriate direction for the company. He formed a new software company called NeXT, Inc. He purchased an animation company the following year, and this company was to become Pixar Animation Studio. It went on to produce highly successful animation films such as Toy Story, and the studio merged with Walt Disney in 2006. This led to Job's becoming the largest shareholder in Walt Disney. NeXT was purchased by Apple in 1997, and Jobs returned to Apple as CEO. Under Job's leadership, Apple became a highly innovative company and introduced products that dazzled its customers. These include products such as the iMac, iBook, the iPod, the iPhone and the iPad.

He met Laurene Powell at Stanford Business School in the early 1990s, and they married in 1991. They had three children and lived in Palo Alto, California. He also had a daughter Lisa from his relationship with Chris Ann Brennan in the late 1970s. The Apple Lisa is named after her.

He was diagnosed with a rare form of pancreatic cancer in 2003. He initially tried alternative medicines, such as herbal remedies, acupuncture and a vegan diet, and delayed appropriate medical intervention and surgery for several months. He then received medical treatment, which appeared to remove the tumour. He took 6 months of medical leave from Apple in 2009 to focus on his health and underwent a successful liver transplant, and his prognosis was described as excellent. He returned to Apple but in early 2011 he went on medical leave again. He resigned as CEO of Apple in August 2011 and died in October 2011.

He received many awards in recognition of his contribution to the computing field. He was a corecipient of the National Medal of Technology with Stephan Wozniak which they received from President Reagan in 1985.

34.1 Apple Computers

Jobs and Wozniak founded Apple Computers in 1976, with Wozniak responsible for product development and Jobs responsible for marketing. The Apple I computer was released in 1977, and it retailed for $666.66. It generated over $700,000 in revenue for the company, but it was mainly of interest to computer hobbyists and engineers. The Apple II computer was released later that year, and this machine included colour graphics, and it came in its own plastic casing. It retailed for $1299, and it was one of the earliest computers to come pre-assembled. It was to generate over $139 million in revenue for the company.

The Apple I computer had 4 K of RAM (expandable to 8 K), 256 bytes of ROM and was intended to be used those for whom computing was a hobby. The user needed to supply a case, power supply and keyboard or display as these were not included.

The Apple II computer was a significant advance on the Apple I, and it was a popular 8-bit home computer. It was one of the earliest computers to have a colour display, and the BASIC programming language was built in. It contained 4 K of RAM (which was could be expanded to 48 K). The VisiCalc spreadsheet program was released on the Apple II, and this helped to transform the computer into a credible business machine. The Apple II was a major commercial success, and Apple became a public listed company in 1980. John Scully became CEO of Apple in 1983 (Fig. 34.2).

The Apple Macintosh was announced during a famous television commercial aired during the Super Bowl in 1984. The Macintosh included a friendly and intuitive graphical user interface (GUI), and the machine was much easier to use than the standard IBM PC. The latter was a command-driven operating system that required the user to be familiar with its commands. The introduction of the Mac

Fig. 34.2 Apple II computer
(Photo Public Domain)

GUI was an important milestone in the computing field, and it demonstrated that the usability of a computer needs to be considered in its design.

Jobs got the idea of a commercial graphical user interface from the work taking place at Xerox's PARC's research centre in Palo Alto in California. They were working on a mouse-driven graphical user interface, and Jobs immediately recognized its potential. Apple intended that the Macintosh computer would be an inexpensive and user-friendly personal computer that would rival the IBM PC and its clones. However, it was more expensive than IBM's personal computer and retailed for $2495. Further, it initially had limited applications available for its customers, whereas the IBM PC had spreadsheets, word processors and databases applications.

The sales of the Macintosh were slow, and Apple went through financial difficulty in the mid-1980s. This led to boardroom tensions at senior management level and an inevitable power struggle between Jobs and Scully. Jobs management style was a little erratic, and he resigned from Apple in 1985.

34.2 NeXT and Pixar

Jobs founded NeXT, Inc., in 1986, and this company produced the NeXT worksta-tion computers. Ross Perot was also an investor in the company. NeXT changed direction to become more software focused, and it released a framework for web application development called WebObjects in 1996. The object-oriented operating system on the NeXT computer (NeXTSTEP) was selected by Apple to be the basis of Apple's next operating system. The Macintosh OS X and iPhone Operating System (*i*OS) were based on this operating system.

Jobs bought a computer graphics company, which he renamed as Pixar, in 1986, and this company went on produce high-quality animation movies such as *Toy Story*.

Disney purchased the company in 2006 for over $7 billion, and Jobs became the largest shareholder in Disney with a 7 % stake in the company.

NeXT was acquired by Apple for over $400 million in 1997, and this led to Jobs return to Apple. He became CEO later that year and was paid an annual salary of $1.[2]

34.3   Apple (Return of Jobs)

Jobs developed an alliance between Apple and Microsoft on his return. The *i*Mac (a Macintosh desktop computer) was released in 1998, and it was a major commercial success for the company. The letter *i* stands for the *Internet* and also represents the fact that the product is a personal device designed for the *individual*. The *i*Mac originally employed the PowerPC chip designed and developed by IBM and Motorola, but these were later replaced with Intel processors (in 2006). The entire Macintosh line was transitioned to Intel processors later that year (Fig. 34.3).

The *i*Book (a line of personal laptops) was introduced in 1999, and the *i*Pod (a portable music player) was introduced in 2001. These products were major commercial successes for Apple. The *i*Pod is a small portable hard disc MP3 player which has a capacity of 5–10 GB, and it can hold up to 1,000 MP3 songs. The *i*Pod prepared the way for *i*Tunes and the *i*Phone.

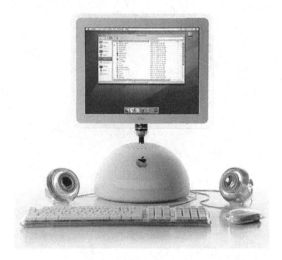

Fig. 34.3 Apple iMac G4
(Photo Public Domain)

---

[2]Jobs joked that $0.50 was for salary and $0.50 was for performance. He was also given generous share options.

The *i*Tunes Music Store was launched in 2003 and is the largest music vendor in the United States. It allows songs to be downloaded for a small fee, and individual songs are sold for the same price and without a subscription fee for access to the catalogue. There are over 20 million songs in the catalogue, and movies are also available for purchase. By 2008 *i*Tunes was the second largest music retailer in the United States.

Apple entered the mobile phone market with the release of the *i*Phone in 2007. This project commenced in 2005, and the resulting Internet-based multimedia smartphone with its touch screen became an immediate success. Its features include a video camera, email, web browsing, text messaging and voice.

It released the *i*Pad in 2010, and this is a large screen tablet-like device. It uses a touch-based operating system.

Jobs suffered from ill health from 2003 when he was diagnosed with a rare type of pancreatic cancer. He fought the disease for several years and died in 2011.

Chapter 35
Gary Kildall

Gary Kildall was an American computer scientist and entrepreneur and the founder of Digital Research, Inc. (DRI). He is famous for his work on developing the first microprocessor disc operating system and for developing the first programming language and compiler for a microprocessor. His CP/M disc operating system was the basis for the operating system used on the IBM personal computer, and if things had turned out differently at the time, his company, Digital Research, Inc., could well have been Microsoft.

Bloomberg Business Week published an article on Kildall in 2004 with the title "The Man Who Could Have Been Bill Gates" [Blo:04]. This article describes the background to the development of an operating system for the IBM PC and the failed negotiations between Digital Research and IBM on the licensing of the CP/M operating system for the IBM PC (Fig. 35.1).

He was born in Seattle, Washington, in 1942. He attended the University of Washington and obtained a bachelor's degree in mathematics in 1967. He became interested in computing and pursued a postgraduate degree in computer science at the university. His Master's degree in computer science was awarded in 1968, and he then embarked on his Ph.D. degree. He became an assistant professor at the Naval Postgraduate School at Monterey in California and became interested in the developments in the computing field at nearby Silicon Valley. His Ph.D. degree in computer science from the University of Washington was awarded in 1972.

Kildall became aware of early work taking place at Intel on microprocessors, and he travelled to Silicon Valley on his days off from the Naval School to work as a consultant with Intel. He soon recognized the potential of the microprocessor as a computer in its own right, and he began writing experimental programs for the newly released Intel 4004. This was the first commercially available microprocessor, and Kildall also worked with Intel on the 8008 and 8080 microprocessors. He developed the first high-level programming language for a microprocessor (PL/M) in 1973, and PL/M allowed programmers to write applications for microprocessors.

He developed the CP/M operating system (Control Program for Microcomputers) in 1973. This operating system allowed the Intel 8080 microprocessor to control a floppy drive, and CP/M combined all of the essential components of a computer

at the microprocessor level. CP/M was the first disc operating system for a microcomputer, and files could be read and written to and from an 8-inch floppy disc. The development of CP/M made it possible for computer hobbyists and companies to build the first personal computers.

He set up Digital Research, Inc., (DRI) with his wife Dorothy in 1976. This California-based company was initially called Galactic Digital Research, Inc., and it was set up to develop, market and sell the CP/M operating system. Kildall made CP/M hardware independent by creating a separate module called the BIOS (*Basic Input/Output System*). He added several utilities such as an editor, debugger and assembler, and by 1977 several manufactures were including CP/M with their systems. There were over a quarter of a million copies of CP/M sold.

IBM approached Digital Research in 1980 with the intention of licensing the CP/M operating system for the new IBM personal computer. However, their negotiations were unsuccessful, and IBM made an agreement with a small company called Microsoft for their new operating system. This operating system was largely based on Kildall's CP/M, and although Kildall considered taking legal action against IBM and Microsoft, he decided against it due to the limited resources of DRI.

He co-hosted a television series called the *Computer Chronicles* for 6 years. The series commenced in 1983 and it followed trends in personal computing. Digital Research was purchased by Novell in 1991 for $120 million, and Kildall became wealthy.

He became embittered as a result of the IBM/Microsoft experience, and in later life he had problems with alcohol addiction. He died in tragic circumstances in 1994 at the young age of 52. He received a posthumous award from the Software Publishers Association (now called the Software and Information Industry Association) in 1995 for his contributions to the microcomputer industry. The University of Washington awards the Gary Kildall Endowment Scholarship annually to outstanding undergraduate computer science students.

35.1 Licensing CP/M to IBM

Kildall lost out on the opportunity of a lifetime to supply the operating for the IBM personal computer to IBM, and instead it was Microsoft that supplied the operating system and reaped the benefits. Microsoft would later become a technology giant and a dominant force in the computer industry.

Don Estridge (discussed in an earlier chapter) led the IBM team that was developing the IBM personal computer. The project was subject to a very aggressive delivery schedule, and while traditionally IBM developed a full proprietary solution, it decided instead to outsource the development of the microprocessor to a small company called Intel and to outsource the development of the operating system. The IBM team initially asked Bill Gates and Microsoft in Seattle to supply them with an operating system. Microsoft had already signed a contract with IBM to supply a BASIC interpreter for the IBM PC, but they lacked the appropriate expertise to develop the operating system. Gates referred IBM to Gary Kildall at DRI, and the IBM team approached Digital Research with a view to licensing the CP/M operating system for their new IBM PC.

Digital Research was working on CP/M-86 for the Intel 16-bit 8086 microprocessor that had been introduced by Intel in 1978. IBM decided to use the Intel 8088 for its new personal computer product, and the 8088 processor, which was introduced in 1979, was a lower cost and slower version of the 8086.

IBM and Digital Research failed to reach an agreement on the licensing of CP/M for the IBM PC. The precise reasons for failure are unclear, but some immediate problems arose with respect to the signing of an IBM non-disclosure agreement during the visit. It is unclear whether Kildall actually met with IBM and whether there was an informal handshake agreement between both parties. There was certainly no legal written agreement between IBM and DRI.

There may also have been difficulties in relation to the amount of royalty payment being demanded by Digital Research, as well as practical difficulties in achieving the required IBM delivery schedule (due to Digital Research's existing commitments to Intel). Kildall was superb at technical innovation, but he may have lacked the appropriate business acumen to secure a good deal.

Gates had been negotiating a Microsoft BASIC license agreement with IBM, and he now saw a business opportunity for Microsoft. He offered to provide an operating system (later called PC-DOS) and BASIC to IBM on favourable terms. The offer was accepted by IBM, and the rest, as they say, is history. Gates was aware of the work done by Tim Patterson on a simple quick and dirty version of CP/M (called QDOS) for the 8086 microprocessor for Seattle Computer Products (SCP). Gates licensed QDOS for $50,000 and hired Patterson to modify it to run on the IBM PC for the 8088 microprocessor. Gates then licensed the operating system to IBM for a low per copy royalty fee. IBM called the new operating system PC-DOS, and Gates retained the rights to MS-DOS which were used on IBM compatible computers produced by other hardware manufacturers. In time, MS-DOS would become the dominant operating system (eclipsing PC-DOS due to the open architecture of the IBM PC and the growth of clones) leading to the growth of Microsoft into a major corporation.

DRI released CP/M-86 shortly after IBM released PC-DOS. Kildall examined PC-DOS, and it was clear to him that it had been derived from CP/M. He was furious and met separately with IBM and Microsoft, but nothing was resolved. Digital Research considered suing Microsoft for copying all of the CP/M system calls in DOS 1.0, as it was evident to Kildall that Patterson's QDOS was a copy of CP/M. He considered his legal options, but his legal advice suggested that as intellectual copyright law had only been recently introduced in the United States, it was not clear what constituted infringement of copyright. There was no guarantee of success in any legal action against IBM, and considerable cost would be involved. Kildall threatened IBM with legal action, and IBM agreed to offer both CP/M-86 and PC-DOS. However, as CP/M was priced at $240 and DOS at $60, few PC owners were willing to pay the extra cost. CP/M was to fade into obscurity.

Perhaps, if Kildall had played his hand differently he could have been in the position that Bill Gates is in today, and Digital Research could have been, in effect the Microsoft of the PC industry. Kildall erred in been slow in developing the 16-bit operating which gave Patterson the opportunity to create his own version. IBM was under time pressure with the development of the IBM PC, and Kildall was unable to meet the IBM deadline. This resulted in IBM dealing with Gates instead of DRI. Further, the royalty fee demanded by Kildall for CP/M was not very clever, as the excessive royalty fee demanded resulted in very low demand for his product, whereas if a sensible price had been charged then DRI may have made some reasonable revenue.

Kildall could justly feel hard done by, and he may have viewed Microsoft's actions as the theft of his intellectual ideas and technical inventions. It shows that technical excellence and innovation is not in itself sufficient for business success, and that a certain business acumen or entrepreneurial flair is also required.

35.2 Computer Chronicles

The Computer Chronicles was an American television series broadcast from 1981 to 2002. The series covered the evolution of the personal computer from its early days, to the rise of the Internet, to the immense market for PCs at the start of the twenty-first century. Kildall served as co-host from 1983 to 1990, and he provided analysis and commentary on computer products and the evolution of the computer industry.

Chapter 36
Donald Knuth

Donald Ervin Knuth is an American computer scientist, and he is known as the *father of the analysis of algorithms*. He has made important contributions to theoretical computer science, to the design of programming languages and to the *art of computer programming* through his series of well-known books. He is the creator of the TeX computer typesetting system (Fig. 36.1).

He was born in 1938 in Wisconsin in the United States and initially studied physics at the Case Institute of Technology. He switched to mathematics in his second year. He became aware of computers in the summer between his first and second year at the university. He was working in the Statistics lab; when he saw the IMB 650 across the hall, he was immediately fascinated, and he later stated that it provided the inspiration for his future work. He learned to program and wrote a user manual and instructional software for the IBM 650. He wrote a number of programs during his undergraduate years including a program to analyze the performance of the institute's basketball team.

He published two scientific papers while still an undergraduate, and he was one of the founding editors of the *Engineering and Science Review*, a technical magazine published by the students of the university. He was recognized as an exceptional student and was awarded a Master of Science degree simultaneously with his Bachelor of Science degree, in 1960.

He obtained a Ph.D. in mathematics from the Californian Institute of Technology in 1963, and the title of his thesis was "Finite Semifields and Projective Planes". He was one of the earliest mathematicians to take computer science seriously, and he introduced rigour and elegance into programming. The computing field at the time was characterized by people writing and rewriting code until all known defects were removed. Knuth's emphasis on rigour aimed to provide a proof that the program was correct. He moved from California Institute of Technology to Stanford University in 1968.

He has made important contributions to the rigorous analysis of the computational complexity of algorithms. He is the author of the influential book *The Art of Computer Programming* which is published in four volumes. This is regarded as the

G. O'Regan, *Giants of Computing: A Compendium of Select, Pivotal Pioneers*,
DOI 10.1007/978-1-4471-5340-5_36, © Springer-Verlag London 2013

Fig. 36.1 Donald Knuth at
Open Content Alliance 2005
(Courtesy of Jacob
Appelbaum)

discipline's definitive reference guide, and it has been a lifelong project for Knuth. The first volume was published in 1968, the second in 1969, the third in 1973, and the fourth appeared in 2011.

He is the creator of the TeX and METAFONT systems for computer typesetting. These programs are used extensively for scientific publishing and were developed for high-quality computerized typesetting. This allows mathematical articles to be printed to a very high quality. The TeX system is the typesetting engine, and METAFONT is the font design system. Knuth's system has been adapted by commercial typesetting systems. He is the author of over 26 books, 160 papers and 5 patents.

He has received numerous awards in recognition of his contributions to the computing field. He was the first recipient of the ACM Grace Murray Hopper Award from the Association of Computing Machinery in 1971. He received the ACM Turing Award in 1974 for major contributions to the analysis of algorithms and the design of programming languages. He received the National Medal of Science award in 1979 and the John von Neumann medal in 1995. He has received numerous honorary doctorates from universities around the world.

He retired early from Stanford University in 1992 to work full time on his books, and he has been professor emeritus of the Art of Computer Programming at Stanford University since 1992.

36.1 Art of Computer Programming

Knuth did some consulting work for Burroughs Corporation in California while doing his Ph.D., and he wrote compilers for various computers. Addison-Wesley approached him in 1962 and asked him to write a textbook on compilers. He signed a book contract intending to write a book that contained 12 chapters.

He began working on this book following the completion of his Ph.D. degree, and as he proceeded with the project, he recognized the need for an organized and reliable book to present the state of the art in the field of computer science. His book grew longer as he wrote it, and when he presented his first draft to Addison-Wesley, it had reached over 2,000 printed pages.

Addison-Wesley decided that the book should be organized into seven volumes, and volume 1 of *The Art of Computer Programming* (TAOCP) was published in 1968. Volumes 2 and 3 were completed by 1973. Volume 4 appeared in 2011.

Volume 1 of this book is a classic in computing programming and presents a wide variety of algorithms and the mathematical analysis of them. It includes the essential mathematical preliminaries for the discipline. Volume 2 covers topics such as random number generators, prime factorization, algorithms for single precision and double precision, arithmetic calculations and algorithms for calculating fractions. Volume 3 is concerned with sorting and searching and examines the existing algorithms to see how they may be improved. Volume 4a considers a selection of algorithms that will still be important in 50 years' time.

These books established the analysis of algorithms as a computer science topic in its own right, and although Knuth suggested to the publishers that the title of the book be renamed to *The Analysis of Algorithms*, it was decided for marketing purposes to stay with the original title.

The collection of techniques, algorithms and theory in these books has served as a focal point for developing curricula and as an organizing influence on computer science.

36.2 TeX

Knuth became interested in typesetting in 1973 when Addison Wesley replaced mechanical typesetting with computerized typesetting. This led to a degradation in the quality of the printing of the new editions of his book series, and this motivated him to explore a new computerized typesetting system. His goal was that it would produce high-quality printed documents, and that it would be independent, as far as possible, of changes in printing technology.

He announced the new system in the 1978 American Mathematical Society Gibbs Lecture entitled "Mathematical typography" [Knu:79]. It was developed with the help of Stanford students and colleagues, and it had three main components. These were the TeX typesetting engine, the METAFONT font design system and the Computer Modern set of type fonts. His approach revolutionized digital typesetting, and Knuth made the source code publicly available. This allowed commercial organizations to adapt his code for commercial typesetting systems. The TeX user group was established in 1980 and is a forum to get user feedback on TeX and to prioritize improvements.

Chapter 37
Gottfried Wilhelm Leibniz

Gottfried Wilhelm Leibniz was a German philosopher, mathematician and inventor in the field of mechanical calculators. He developed the binary number system used in digital computers and invented the calculus independently of Sir Isaac Newton. He was embroiled in a bitter dispute towards the end of his life with Newton, as to who developed the calculus first (Fig. 37.1).

He was born in 1646 in Leipzig, Germany, to a wealthy Lutheran family, and his father was a professor of moral theology. He pursued a bachelor's degree at the University of Leipzig from 1661 to 1663, and his studies included philosophy, mathematics, Latin, Greek and rhetoric. He obtained a Master's degree in philosophy from the university in 1663, and his thesis combined aspects of philosophy and law. He completed his Doctorate in Law at the University of Altdorf in 1667.

He was employed by Baron Johann Christian von Boineburg from 1667 until 1676, when he accepted a position with the Duke of Hanover, Johann Friedrich. He was to remain with the House of Hanover until his death in 1716. He acted an assistant, librarian, advisor, lawyer and friend and was involved in a number of scientific, political and literary projects with the House of Hanover. He also wrote a number of monographs on religious topics and on points at issue between the churches.

His interests began to move in a scientific direction from the 1670s. He spent several years in Paris and made contact with some of the leading intellectual figures of the seventeenth century including philosophers and scientists. He studied mathematics and physics under the direction of the physicist Christopher Huygens, and his flair for mathematics soon became apparent. He invented infinitesimal calculus in the late 1670s independently of Newton, and his publication on calculus appeared in 1684. This paper introduced Leibniz's notation on calculus as well as the rules for differentiation of powers, products and quotients. It includes the familiar integral and differential notation that is used today such as the dy/dx notation and the integral notation $\int f(x)dx$.

Newton's publication on calculus appeared some time after Leibniz's, and this led to a bitter controversy between them, with the former accusing Leibniz of plagiarizing his work. Today, it is accepted that both of them independently invented

Fig. 37.1 Gottfried Wilhelm
Leibniz

the calculus. Leibniz's notation in calculus is more popular than Newton's. The last years of Leibniz's life were spent embroiled in this controversy.

He began construction of a calculating machine in the 1670s, and he demonstrated this incomplete machine to the Royal Society in London in 1673. He was elected a fellow of the Royal Society later that year. He experimented with machines to drain water from the mines in the Harz Mountains, and his idea was to use wind power and water power to operate pumps.

Leibniz developed the binary system[1] of arithmetic, and binary numbers are at the heart of modern digital computers. The binary number system uses two symbols 0 and 1, and *a binary digit may be represented by a switch that is either off or on*. However, his calculating machine used decimal numbers rather than binary, and Boole (discussed in an earlier chapter) later used the binary system to develop a calculus of logic that was used by Shannon as the basis for digital computing.

Leibniz was a philosopher and he did important work in philosophy. He developed the theory of monads, and he believed that there were an infinite number of monads. He presented several arguments for the existence of God, and these include the *ontological argument* and the *cosmological argument*. He argued that this world, though not perfect, is the best of all possible worlds, and that the presence of evil in the world (*theodicy*) is not an argument against the goodness of God.

37.1   Step Reckoner Calculating Machine

Leibniz became familiar with Pascal's calculating machine, the *Pascaline*, while in Paris in the early 1670s. He recognized the limitations of the machine as it was capable of performing addition and subtraction operations only. He became

---

[1]The Indian scholar Pinggala did some early work on a version of the binary number system about 2,000 years before Leibniz.

Fig. 37.2 Replica of Step Reckoner at Technische Sammlungen Museum, Dresden

interested in the problem of extending the machine for multiplication and division and for speeding up calculation.

He designed and developed a calculating machine that could perform addition, subtraction, multiplication, division and the extraction of roots. He commenced work on the machine in 1672, and the machine was completed in 1694. It was the first calculator that could perform all four arithmetic operations and was superior to the existing Pascaline machine.

Leibniz's machine was called the *Step Reckoner*, and it allowed the common arithmetic operations to be carried out mechanically. A metallic version of the machine was presented to the French Academy of Sciences in 1675. He visited the Royal Society in London in 1673 to demonstrate a prototype of the incomplete machine, and this gave him the opportunity to meet leading scientists such as Hooke and Boyle (Fig. 37.2).

The operating mechanism used in his calculating machine was based on a counting device called the stepped cylinder or "Leibniz wheel". This mechanism allowed a gear to represent a single decimal digit from zero to nine in just one revolution, and this mechanism remained the dominant approach in the design of calculating machines for the next 200 years. The Leibniz wheel was essentially a counting device consisting of a set of wheels that were used in calculation. The Step Reckoner consisted of an accumulator which could hold 16 decimal digits and an 8-digit input section. The eight dials at the front of the machine set the operand number which was then employed in the calculation.

The machine performs multiplication by repeated addition and division by repeated subtraction. The basic operation is to add or subtract the operand from the accumulator as many times as desired. The machine could add or subtract an 8-digit number to the 16-digit accumulator to form a 16-digit result. It could multiply two 8-digit numbers to give a 16-digit result, and it could divide a 16-bit number by an 8-digit number. Addition and subtraction is performed in a single step, with the operating crank turned in the opposite direction for subtraction. The result is stored in the accumulator.

37.2 Binary Numbers

Arithmetic has traditionally been done using the decimal notation,[2] and this involves using the digits 0, 1, 2, ... 9, and then 10, 11, ... 19 and so on. Leibniz was one of the earliest people to recognize the potential of the binary number system, and this system uses just two digits namely "0" and "1". The number two is represented by 10, the number four by 100 and so on. Leibniz described the binary system in *Explication de l'Arithmétique Binaire* [Lei:03] which was published in 1703. A table of values for the first 15 binary numbers is given in Table 37.1.

Leibniz's 1703 paper describes how binary numbers may be added, subtracted, multiplied and divided, and he was an advocate of their use. The key advantage of the use of binary notation is in digital computers, where a binary digit can be implemented by an *on-off switch*, with the digit 1 representing that the switch is on, and the digit 0 representing that the switch is off. Claude Shannon (discussed in a later chapter) showed in his Master's thesis [Sha:37] that the binary digits (i.e. 0 and 1) can be represented by electrical switches. This allows binary arithmetic and more complex mathematical operations to be performed by relay circuits and provides the foundation of digital computing.

Table 37.1 Binary number system

Binary	Dec.	Binary	Dec.	Binary	Dec.	Binary	Dec.
0000	0	0100	4	1000	8	1100	12
0001	1	0101	5	1001	9	1101	13
0010	2	0110	6	1010	10	1110	14
0011	3	0111	7	1011	11	1111	15

[2]Other bases have been employed such as the segadecimal (or base-60) system employed by the Babylonians. The decimal system was developed by Indian and Arabic mathematicians between 800 and 900 AD and was introduced to Europe in the late twelfth/early thirteenth century. It is known as the *Hindu-Arabic system*.

Fig. 37.3 Derivative as the tangent to curve

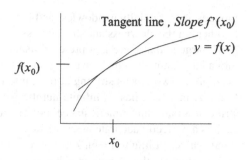

37.3   Differential and Integral Calculus

Differential calculus is the mathematics of motion and acceleration, whereas integral calculus is applicable to problems concerning area and volume. Calculus is a major branch of mathematics and was developed independently by Leibniz and Newton. The question as to who first developed the differential and integral calculus led to a major controversy between Newton and Leibniz, and the consensus today is that both Newton and Leibniz independently developed the calculus. In fact, today it is Leibniz's notation that is more widely used.

They developed the underlying theory of calculus in the late seventeenth century, and it became clear that calculus may be applied to practical problems in physics. Leibniz's notation includes the well-known dy/dx notation used in differential calculus and the notation $\int f(x)dx$ used in integral calculus (Fig. 37.3).

Leibniz commenced his work on calculus in 1673, and he visited England in 1673 and 1676. He published his work on the calculus in 1684 with his publication *Nova Methodus*. Newton commenced his work on around 1666 and had already developed a theory of tangents around the time that Leibniz had started his work on the calculus. Much of the controversy centred on whether Leibniz had seen some of Newton's unpublished manuscripts during his visits to England.

Calculus has many applications in science and astronomy. Differential calculus is applicable to problems involving motion and change and to determining the tangent to a curve. Integral calculus is applicable to problems involving the computation of area and volume.

37.4   Philosophy

Leibniz's philosophy is based on the idea of substance, and he believed that there is an infinite number of substances which he termed *monads* [Rus:45]. Each monad is essentially a soul, and Leibniz rejected the reality of matter and therefore seemed to believe instead in an infinite family of souls. He was of the view that no two monads could have a causal relationship with each other, and that any appearance of a causal relationship is deceptive. He believed that every monad mirrors the universe and that monads are essentially windowless and form a hierarchy, in which some are superior to others in the clearness and distinctness to which they mirror the universe.

He argues that there is a pre-established harmony between the changes in one monad and another, which gives the impression of interaction. This is similar to the example of two clocks striking at the same time, as they are keeping perfect time. Leibniz, in effect, has an infinite number of clocks; all arranged by the creator to strike at a particular time instance, not because they affect each other but because each is a perfectly accurate mechanism.

Space, according to Leibniz, is not real, but it has a real counterpart, namely, the arrangements of the monads in a three-dimensional order. Leibniz had a principle of sufficient reason, according to which nothing happens without a reason. He argues that God always acts for the best although he is under no logical necessity to do so.

Leibniz formulated various arguments for the existence of God. Many of these arguments existed in various forms Aristotle, St. Anselm, St. Thomas Aquinas and Descartes. Leibniz stated these arguments better than they had been stated previously.

Chapter 38
Ada Lovelace

Lady Augusta Ada Byron was an English mathematician who collaborated with Babbage on applications for the analytic engine. She is considered the world's first programmer, and the Ada programming language is named in her honour (Fig. 38.1).

She was born in London in 1815 and was the only legitimate daughter[1] of the English poet, Lord Byron. Her parents separated shortly after her birth, and her mother, Annabella Milbanke, who was a mathematician, was granted sole custody of her. Lady Byron arranged for Ada to be educated in science and mathematics by private tutors. Lord Byron left England in 1816 never to return, and he died in Athens in 1824 when Ada was just nine. She never met her father but is buried beside him in the Church of St. Mary Magdalene, Hucknall, in Nottinghamshire.

She was presented at court in 1833 and introduced to Babbage at a dinner party later that year. She and her mother visited Babbage's studio in London, where the prototype Difference Engine was on display. The machine fascinated Ada, and she recognized the beauty of the invention.

She became friends with the British mathematician and astronomer, Mary Somerville, in 1834. Somerville guided Ada in her study of mathematics and sent her mathematics books and problems to solve. They spoke regularly on mathematics and science, and this included discussion on Babbage's calculating machines. Ada was fascinated by the idea of the analytic engine, and she communicated regularly with Babbage with ideas on its applications.

She married William King (who became the Earl of Lovelace in 1838) in 1835, and she became the Countess of Lovelace. They had three children: Byron,

---

[1] It is likely that Lord Byron was the father of Elizabeth Medora Leigh as he is believed to have had an incestuous affair with his half-sister, Augusta Leigh. Annabella Milbanke told Lovelace that Leigh was her half-sister and fathered by Byron. He was also the father of Clara Allegra Byron, the illegitimate daughter of Claire Clairmont. She died of typhus aged five during Byron's travels in Italy.

G. O'Regan, *Giants of Computing: A Compendium of Select, Pivotal Pioneers*,
DOI 10.1007/978-1-4471-5340-5_38, © Springer-Verlag London 2013

Fig. 38.1 Lady Ada
Lovelace

Annabella and Ralph Gordon. Ada began studies in advanced mathematics with the British mathematician and logician Augustus De Morgan, in 1841.

She produced an annotated translation of Menabrea's *Notions sur la machine analytique de Charles Babbage*. She explained in the notes how the Analytic Engine could be programmed and provided what is considered to be the *first computer program*. This program provided a written plan for how the analytic engine would calculate *Bernoulli numbers. She is therefore considered to be the first computer programmer.*[2]

The computer programming language developed in the late 1970s by the US Department of Defense was named *Ada* in her honour. The Department of Defense Military Standard for the language is MIL-STD-1815, where 1815 is the year of her birth. The British Computer Society awards the *Lovelace medal* to individuals who have made an outstanding contribution to the understanding or advancement of computing, and the winner is invited to give the BCS public *Lovelace lecture* the following year. Her achievements remain an inspiration to women in science, engineering and mathematics. There is an annual "Ada Lovelace Day" held in October, and its goal is to raise the profile of women in mathematics, science and engineering.

She suffered from illness throughout her life and died of uterine cancer at the young age of 37 in 1852.

38.1 Applications of Analytic Engine

Babbage (discussed in an earlier chapter) intended that the operation of the Analytic Engine would be analogous to the operation of the Jacquard loom. The latter is capable of weaving (i.e. executing on the loom) a design pattern that has been

---

[2]However, clearly Babbage would have produced several plans for calculations before her.

prepared by a team of skilled artists. The design pattern is represented by a set of cards with punched holes on each card, and each card represents a row in the design. The cards are then ordered and placed in the Jacquard loom, and the loom produces the exact pattern.

The use of punched cards in the Analytic Engine allows the formulae to be manipulated in a manner dictated by the programmer. The cards command the analytic engine to perform various operations and to return a result.

The Analytic Engine was designed in 1834 as the world's first mechanical computer [Bab:42]. It included a processor, memory and a way to input information and output results. Babbage intended that the program be stored on read-only memory using punched cards, and that the input and output would be carried out using punched cards. He intended that the machine would be able to store numbers and intermediate results in memory that could then be processed. There would be several punched card readers in the machine for programs and data. He envisioned that the machine would be able to perform conditional jumps as well as parallel processing where several calculations could be performed at once.

However, the machine was never built due to funding issues. Babbage had already received funding from the British government to build his *difference engine* (a machine to compute tables of values of polynomial functions), but he had produced only an early prototype of the machine. Therefore, the British government was unwilling to commit further funds as his earlier project was incomplete.

Lovelace remained in regular correspondence with Babbage, and she produced an annotated translation of Menabrea's *Notions sur la machine analytique de Charles Babbage* [Lov:42]. The notes that she added to the translation were about three times the length of the original memoir and considered many of the difficult and abstract questions connected with the subject. These notes are regarded as a description of a computer and software, and they were published in Richard Taylor's Scientific Memoirs (vol. 3) in 1843. Her notes on the analytic engine were republished in 1953.

She explained in the notes how the Analytic Engine could be programmed and wrote what is considered to be the first computer program. This program detailed a plan be written for how the engine would calculate *Bernoulli numbers*. Lady Ada Lovelace is therefore considered to be the first computer programmer. She was called the *enchantress of numbers* by Babbage.

She saw the potential of the analytic engine to fields other than mathematics. She predicted that the machine could be used to compose music, produce graphics, as well as solving mathematical and scientific problems. She speculated that the machine might act on other things apart from numbers and be able to manipulate symbols according to rules. In this way, a number could represent an entity other than a quantity, and she believed the analytic engine had applications outside of mathematics.

She suffered from health problems and had issues with alcohol addiction and gambling later in life. She died of cancer in 1852.

Chapter 39
John McCarthy

John McCarthy was an American computer scientist and the father of the artificial intelligence (AI) field. The term *artificial intelligence* was coined by McCarthy in 1955, and he was one of the founders of the field. He developed the Lisp programming language, which is one of the oldest programming languages and remains a popular language in the AI field. He served on the international committee that developed the influential Algol programming language. He did some early work on chess playing programs, and he participated in an international match against rivals in Russia conducted via telegraph. He developed the important concept of time-sharing computer systems in the late 1950s/early 1960s. He has also worked on proving that computer programs meet their specifications (Fig. 39.1).

He was born to immigrant parents in Boston, Massachusetts, in 1927, and his family relocated to California during the Great Depression. He obtained a bachelor's degree in mathematics from California Institute of Technology (Caltech) in 1948. He earned his Ph.D. in mathematics from Princeton University in 1951. He held a number of appointments at Princeton, Dartmouth and MIT before he became a professor at Stanford in 1962 where he remained until his retirement in 2000. He founded the AI laboratory at MIT with Marvin Minsky in 1957, and he later set up the AI laboratory at Stanford.

McCarthy advocated the use of mathematical logic in the AI field, and he did extensive work on the formalization of common-sense knowledge and reasoning. He believed that mathematical logic could formalize knowledge and reasoning, and he developed a program that could draw conclusions from a set of premises. He considered a program to have common sense if it could deduce relevant conclusions from what it already knows.

He has received several awards in recognition of his contributions to the computing field. He received the ACM Turing Award in 1971 for his contribution to the AI field. He received the National Medal of Science in mathematical, statistical and computational sciences in 1990. He received the Kyoto Prize of the Inamori Foundation in 1988, and he died in 2011.

G. O'Regan, *Giants of Computing: A Compendium of Select, Pivotal Pioneers*, DOI 10.1007/978-1-4471-5340-5_39, © Springer-Verlag London 2013

Fig. 39.1   John McCarthy

39.1   Artificial Intelligence

McCarthy coined the term "artificial intelligence" in 1955. The term appeared in the proposal for the Dartmouth Summer Research Project on Artificial Intelligence, which was written by McCarthy and others.

The goals of the early pioneers of artificial intelligence were to achieve human-level AI. The success of early AI went to its practitioners' heads, and they believed that they would soon develop machines that would emulate human intelligence. They had some initial (limited) success with machine translation, pattern recognition and automated reasoning. However, it is evident today that AI is a long-term project.[1]

McCarthy proposed a program called the Advice Taker in his influential paper "Programs with common sense" [Mc:59]. His idea was that this program would be able to draw conclusions from a set of premises, and McCarthy states that a program has common sense if it is capable of automatically deducing for itself *a sufficiently wide class of immediate consequences of anything it is told from what it already knows*. This paper was the inspiration for further work on question-answering systems and *logic programming*.

The Advice Taker uses logic to represent knowledge (i.e. premises that are taken to be true), and it then applies the deductive method to deduce further truths from the relevant premises.[2] That is, the program manipulates the formal language (e.g. predicate logic) and provides a conclusion that may be a *statement* or an *imperative*. McCarthy's goal was that the program would be able to learn from its experience as effectively as humans do and to improve. The program will have all the logical consequences of what it has already been told and its existing knowledge.

---

[1]This may be hundreds or thousands of years.

[2]Of course, the machine would somehow need to know what premises are relevant and should be selected for to apply the deductive method from the many premises that are already encoded.

His philosophy is that common-sense knowledge and reasoning can be formalized with logic. A particular system is described by a set of sentences in logic. These logic sentences represent all that is known about the world in general and what is known about the particular situation, as well as the goals of the system. The program then performs actions that it infers are appropriate for achieving its goals. That is, common-sense[3] knowledge is formalized by logic, and logical reasoning solves common-sense problems.

The formalization requires sufficient understanding of the common-sense world, and often the relevant facts to solve a particular problem are unknown. It may be that knowledge thought to be relevant is irrelevant and vice versa. A computer may have millions of facts stored in its memory, and the problem is how to determine which facts are relevant from its memory, to serve as the premises in logical deduction.

McCarthy's (1959) paper discusses various common-sense problems such as getting home from the airport. Other examples are diagnosis, spatial reasoning and understanding narratives that include temporal events. Mathematical logic is the standard approach to express premises, and it includes rules of inferences that are used to deduce valid conclusions from a set of premises. Deductive reasoning is a rigorous way of showing how new formulae may be logically deduced from an existing set or premises.

McCarthy's approach to programs with common sense has been criticized by Bar-Hillel and others on the grounds that *common sense is fairly elusive*, and the difficulty that a machine would have in *determining which facts are relevant to a particular deduction from its known set of facts*. However, logic remains an important area in AI.

McCarthy invented the Lisp programming language (list processing) in the late 1950s, while he was based at the Massachusetts Institute of Technology [Mc:60]. It is one of the oldest programming languages and is still used today in the AI field. It is an expression-oriented language, and there is no distinction made between expressions and statements. All code and data are written as expressions, and when an expression is evaluated, it produces a value that may then be used in another expression.

McCarthy introduced two types of expression in his 1958 paper, but only the S-expressions (symbolic expressions), which mirror the representation of code and data, became widely used. The M-expressions (meta-expressions), which express functions of S-expressions, never became popular. Lisp differs from other programming languages in its wide use of parentheses. Lisp functions are written as lists and are processed exactly like data.

---

[3]Common sense includes basic facts about events, beliefs, actions, knowledge and desires. It also includes basic facts about objects and their properties.

Chapter 40
John Mauchly

John Mauchly was an American physicist and engineer who made important contributions to the computing field. He is famous for his work on the design and development of the ENIAC computer, which was one of the earliest digital computers. He was also involved in the design of the EDVAC computer, and he later set up the Eckert-Mauchly Computer Corporation with Presper Eckert. This company pioneered some fundamental computer concepts and later developed the UNIVAC computer (Fig. 40.1).

He was born in Ohio in 1907 and studied electrical engineering for his undergraduate degree at John Hopkins University. He earned a Ph.D. in physics in 1932 from the same university and became a professor of physics at Ursinus College in Philadelphia. He taught there until 1941, when he joined the lecturing staff at the Moore School of Electrical Engineering, at the University of Pennsylvania.

He met Presper Eckert, who was an engineering student, at the Moore School, and they later went on to design and build the ENIAC, EDVAC and the UNIVAC 1. He made a proposal to build an electronic computer using vacuum tubes that would be much faster and more accurate than the existing differential analyzer in the school. He discussed his idea further with representatives with the US Army, and they agreed to provide the funding to build the machine in 1943.

This machine was called ENIAC, and it was one of the earliest digital computers. Mauchly focused on the design of the machine, and Eckert on the hardware engineering side. They were aware of the limitations of ENIAC and commenced work on the design of a successor computer, EDVAC, in early 1945.

von Neumann became involved in some of the engineering discussions, during the development of EDVAC, and he produced a draft report describing this computer. This report was intended to be internal, but circumstances changed due to legal issues which arose with respect to intellectual property and patents. This led to the resignation of Mauchly and Eckert from the Moore School, as they wished to protect their patents on ENIAC and EDVAC. They set up their own company to exploit the new computer technology.

G. O'Regan, *Giants of Computing: A Compendium of Select, Pivotal Pioneers*,
DOI 10.1007/978-1-4471-5340-5_40, © Springer-Verlag London 2013

Fig. 40.1 John Mauchly

The Moore School then removed the names of Mauchly and Eckert from the report and circulated the report to the wider community. The report mentioned the fundamental computer architecture that is known today as the *von Neumann architecture*, and Mauchly and Eckert received no acknowledgement for their contribution.

Eckert and Mauchly's company was called the Eckert-Mauchly Computer Corporation (EMCC), and it was set up in 1947. They received an order from the National Bureau of Standards to develop the Universal Automatic Computer (UNIVAC), which was one of the first commercially available computers. It was delivered in the late 1950s and was used for general processing. EMCC was one of the earliest computer companies, and it pioneered some fundamental computer concepts such as the *stored program*, *subroutines* and *programming languages*. It later developed the UNIVAC 1 and BINAC, before it was taken over by Remington Rand in 1950. This company became Sperry Rand in 1955.

Mauchly founded the consulting company, Mauchly Associates, in 1959. He became embroiled in a legal dispute in the 1973 *Honeywell vs. Sperry Rand* patent court case in the United States. This legal controversy arose from a patent dispute between Sperry and Honeywell, and John Atanasoff (discussed in an earlier chapter) was called as an expert witness in the case. Atanasoff's ABC computer was ruled by the court to be the first electronic digital computer, and the legal judgement confirmed that the ABC existed as *prior art* at the time of Mauchly and Eckert's patent application. It is fundamental in patent law that the invention is novel, and that there is no existing prior art. This meant that Mauchly and Eckert's patent application was invalid, and John Atanasoff was named by the US court as the inventor of the first digital computer. The court ruled that Mauchly derived his invention of ENIAC from Atanasoff. Mauchly died in 1980.

40.1 ENIAC

The Electronic Numerical Integrator and Computer (ENIAC) was one of the first large general-purpose electronic digital computers. It was used to integrate ballistic equations and to calculate the trajectories of naval shells. It was completed in 1946 and remained in use until 1955. The original cost of the machine was approximately $500,000 (Fig. 40.2).

The machine had to be physically rewired in order to perform different tasks, and it was clear that there was a need for an architecture that would allow a machine to perform different tasks without physical rewiring each time. This eventually led to the concept of the *stored program*, which was implemented in the successor to ENIAC. The idea of a stored program is that the program is stored in memory, and when there is a need to change the task that is to be computed, then all that is required is to place a new program in memory rather than rewiring the machine. EDVAC (the successor of ENIAC) implemented the concept of a stored program in 1949, just after its implementation on the Manchester Baby prototype machine in England. The concept of a stored program and von Neumann architecture is detailed in von Neumann's report on EDVAC [VN:45].

ENIAC was a large bulky machine and it was over 100-ft long, 10-ft high, 3-ft deep and weighed about 30 tonnes. Its development commenced in 1943 at the University of Pennsylvania, and it was built for the US Army's Ballistics Research Laboratory. The project team included Presper Eckert as chief engineer, John Mauchly as a consultant and several others. It generated a vast quantity of heat, since each vacuum tube generated heat like a light bulb, and there were over 18,000

Fig. 40.2 Setting the switches on ENIAC's function tables (US Army photo)

Fig. 40.3 Replacing a valve on ENIAC (US Army photo)

tubes in the machine. The machine employed 150 kW of power and air-conditioning was used to cool it.

It employed decimal numerals and it could add 5,000 numbers and do 357 10-digit multiplications or 35 10-digit divisions in 1 s. It could be programmed to perform complex sequences of operations, and this included loops, branches and subroutines. However, the task of taking a problem and mapping it onto the machine was complex, and it usually took weeks to perform. The first step was to determine what the program was to do on paper; the second step was the process of manipulating the switches and cables to enter the program into ENIAC, and this usually took several days. The final step was verification and debugging, and this often involved single-step execution of the machine.

There were problems initially with the reliability of ENIAC as several vacuum tubes burned out most days. This meant that the machine was often nonfunctional as high-reliability tubes were not available until the late 1940s. However, most of these problems occurred during the warm-up and cool-down periods, and, therefore, it was decided not to turn the machine off. This led to improvements in its reliability to the acceptable level of one tube every 2 days. The longest continuous period of operation without a failure was 5 days (Fig. 40.3).

The very first program run on ENIAC took just 20 s, and the answer was manually verified to be correct after 40 h of work with a mechanical calculator. One of the earliest problems solved was related to the feasibility of the hydrogen bomb. It involved the input of 500,000 punched cards, and the program ran for 6 weeks and gave an affirmative reply. ENIAC was preceded in development by the Atanasoff Berry Computer (ABC) and the Colossus computer in the UK. ENIAC was a major milestone in the development of computing.

40.2   EDVAC

The EDVAC (Electronic Discrete Variable Automatic Computer) was the successor to the ENIAC. It was a stored-program computer and it cost $500,000. It was proposed by Eckert and Mauchly in 1944, and design work commenced prior to the completion of ENIAC.

It was delivered to the Ballistics Research Laboratory in 1949, and it commenced operations in 1951. It remained in operations until 1961. It employed 6,000 vacuum tubes and its power consumption was 56,000 W. It had 5.5 kB of memory.

The EDVAC was one of the earliest stored-program computers, and it employed the program instructions in memory, rather than rewiring the machine each time.

40.3   Controversy: ABC and ENIAC

The ABC computer was ruled to be the first electronic digital computer in the 1973 *Honeywell vs. Sperry Rand* patent court case in the United States. The court case arose from a patent dispute between Sperry and Honeywell, and Atanasoff was called as an expert witness in the case.

The court ruled that Eckert and Mauchly did not invent the first electronic computer, since the ABC existed as *prior art* at the time of their patent application. It is fundamental in patent law that an invention is novel, and that there is no existing prior art. This meant that the Mauchly and Eckert patent application for ENIAC was invalid, and Atanasoff was named as the inventor of the first digital computer.

Mauchly had visited Atanasoff on several occasions prior to the development of ENIAC, and they had discussed the implementation of the ABC. Mauchly subsequently designed the ENIAC, EDVAC and UNIVAC. The court ruled that the ABC was the first digital computer and that the inventors of ENIAC had derived the subject matter of the electronic digital computer from Atanasoff.

Chapter 41
Marvin Minsky

Marvin Minsky is an American cognitive scientist, a pioneer of robotics and neural networks, author, inventor and one of the founders of the artificial intelligence field. He is Toshiba professor of media arts and sciences and professor of electrical engineering and computer science at the Massachusetts Institute of Technology.

He has made important contributions to artificial intelligence, especially on learning, knowledge representation, common-sense reasoning, neural networks, computer vision and robot manipulation. He has also made important contributions to cognitive psychology, automata theory and symbolic mathematics (Fig. 41.1).

He was born in New York in 1927 and served in the US Navy from 1944 to 1945. He obtained a bachelor's degree in mathematics from Harvard in 1950, and earned his Ph.D. in mathematics from Princeton University in 1954. *His Ph.D. thesis was the first publication on theories and theorems about learning in neural networks* and covered reinforcement and synaptic modifications.

He joined the staff of Massachusetts Institute of Technology in 1959 and has spent his working life at the institute. He was one of the co-founders (along with John McCarthy) of Massachusetts Institute of Technology's Artificial Intelligence laboratory in the late 1950s.

His many inventions include the first neural network simulator (SNARC) created in 1951. He invented the first head-mounted graphic display in 1963 and a hydraulic robot arm in 1967. He is the author of several books on the AI field including the *Society of Mind* which was published in 1985.

He received the ACM Turing Award in 1969 in recognition of his central role in creating, shaping, promoting and advancing the field of artificial intelligence. He received the Computer Pioneer Award from IEEE Computer Society in 1995.

G. O'Regan, *Giants of Computing: A Compendium of Select, Pivotal Pioneers*,
DOI 10.1007/978-1-4471-5340-5_41, © Springer-Verlag London 2013

Fig. 41.1 Marvin Minsky

41.1   Artificial Intelligence

Minsky was one of the founders (with John McCarthy) of the AI field. The goals of the early pioneers of artificial intelligence were to achieve artificial human-level intelligence, but it is evident today that AI is a long-term project. Minsky has worked in many areas of the AI field, including neural networks, robotics, learning and knowledge representation.

He built the first randomly wired neural network learning machine (SNARC – *Stochastic Neural Analog Reinforcement Computer*) in 1951. He invented the confocal microscope in 1957, which overcame limitations of conventional microscopes and significantly enhanced resolution and image quality.

His symbol manipulation approach to AI dates from the early days of the AI field. His 1961 paper, "Steps towards artificial intelligence" [Min:61], examined the current state of the field and outlined key problems that needed to be solved. The paper considered the problems of *heuristic programming* and making computers solve really difficult problems. The problems are divided into five main areas, namely, heuristic search, pattern recognition, learning, planning and induction. The paper *established symbol manipulation at the centre of any attempt to understand intelligence*.

His 1965 paper "Matter, mind and models" [Min:65] is concerned with the problem of making intelligent machines that have self-awareness. The paper considers the problem of acquiring knowledge about the world and the extent to which a model may act as a representation of knowledge. He considered the problems of *dualism* and *free will* and suggested that when intelligent machines are built, they will be as confused as mankind on the philosophical problems *of mind-matter, consciousness* and *free will*.

Minsky and John McCarthy founded the MIT AI lab in the late 1950s, and Minsky served as codirector from 1959 to 1974. The goals of the lab were to gain an improved understanding of intelligence and to create machines that would exhibit a degree of intelligence. The lab attempted to model human perception and intelligence and attempted to build practical robots. Minsky designed and built mechanical hands and an arm with 14 degrees of freedom. He argued that space exploration, nuclear safety and so on would be vastly simpler with manipulators driven locally by intelligent computers or remotely by human operators.

He invented the first head-mounted graphical display in 1963 and developed an educational robot (the *Logo Turtle* robot) with Seymour Papert in 1969.

Minsky began to do some work on *perceptrons* in the late 1960s. These are simple computational devices that capture some of the characteristics of neural behaviour (a type of artificial neural network). They were originally developed by Frank Rosenblatt, and Minsky and Papert identified their strengths and limitation.

Minsky and Papert published the book *Perceptrons: An Introduction to Computational Geometry* [MiP:69] in 1969, and this influential but controversial book had an immediate impact on the AI field. It contained several mathematical proofs relating to perceptrons, and it highlighted their strengths and limitations (specifically in relation to the computation of some predicates). These limitations had an influence on the future direction of AI research, which moved towards symbolic systems.

Minsky and Papert continued their collaboration on theories of intelligence and new approaches to childhood education using Logo (the educational programming language developed by Papert). Minsky developed the theory of frames in the mid-1970s, and he emphasized two key concepts in his famous paper *A Framework for Representing Knowledge* [Min:74]. His core argument in this paper is that whenever one encounters a new situation, one selects from memory a structure termed a *Frame*. This is a remembered framework to be adapted to fit reality by changing details as necessary. He argued that:

– Objects and situations may be represented by sets of slots and slot-filling values.
– Many slots can be filled by inheritance from the default descriptions embedded in a class hierarchy.

For example, the frame describing a birthday party would have a slot for the person whose birthday it is, the person's age, location and so on. The theory of frames had an impact on the AI field and especially in the area of expert systems. Minsky investigated knowledge and memory in *A Theory of Memory* [Min:79] and considered the question of how information is represented as well as how it is stored, retrieved and used.

He published *The Society of Mind* in 1985 [Min:88], and this book is concerned with the nature of mind and intelligence. He published *The Emotive Machine* in 2006, and this book is concerned with the nature of consciousness, emotions and common sense.

Chapter 42
Gordon Moore

Gordon Moore is an American computer scientist, entrepreneur and philanthropist. He was a co-founder (with Robert Noyce) of Intel Corporation in 1968, and he served initially as the executive vice president of the company. He was chief executive officer (CEO) of Intel from 1975 to 1987. He has made important contributions to the semiconductor field and is famous for his articulation of *Moore's law* in 1965. His initial formulation of the law predicted that the number of transistors that could be placed on a computer chip (i.e. the transistor density) would double every year. He revised his law in 1975 to state that the transistor density will double roughly every 2 years. His law has proved to be quite accurate, as the semiconductor industry has developed more and more powerful chips at lower costs (Fig. 42.1).

He was born in San Francisco, California, in 1929. He studied chemistry at the University of California, Berkeley, and obtained a bachelor in science from the university in 1950. He then pursued postgraduate studies at the California Institute of Technology (Caltech) and earned a Ph.D. in chemistry in 1954. He completed his postdoctoral work at the John Hopkins University Applied Physics Laboratory from 1954 to 1956.

He then joined Shockley Semiconductor Laboratory (part of Beckmann Instruments), which had been founded by William Shockley. Shockley (discussed in a later chapter) was one of the inventors of the transistor at Bell Labs, but he was not an easy person to work with. Following internal disagreements at the company, eight employees (including Moore) left to set up Fairchild Semiconductors. The group of eight included Robert Noyce who would later co-found Intel with Moore.

Germanium was the most common material for making semiconductors at the time, but Noyce proposed that silicon be employed to make silicon semiconductors. Fairchild pioneered the use of the *planar process* for making transistors, and this process was soon employed by the existing semiconductor companies. Jack Kirby of Texas Instruments succeeded in building an integrated circuit made of germanium containing several transistors in 1958. Noyce built an integrated circuit on a single wafer of silicon in 1960.

G. O'Regan, *Giants of Computing: A Compendium of Select, Pivotal Pioneers*,
DOI 10.1007/978-1-4471-5340-5_42, © Springer-Verlag London 2013

Fig. 42.1 Gordon Moore
(Courtesy of Steve Jurvetson)

Moore was the Director of the Research and Development Laboratories at Fairchild, and he formulated his famous *Moore's law* in 1965 [Mor:65]. The law predicted that the number of transistors on an integrated circuit would double approximately every year. Moore predicted that this law would remain valid for many years, and he revised the law in 1975 to state that the transistor density would double every 2 years.

Noyce and Moore resigned from Fairchild Semiconductors in 1968 and set up Intel the same year. *Intel introduced the world's first microprocessor, the Intel 4004*, in 1971, and it became the industry leader in the microprocessor field. Intel has developed innovative products such as the 4004 microprocessor, the 8-bit 8008 microprocessor, the 8080 microprocessor with about 4,5000 transistors, the 8085 microprocessor, the 16-bit 8086 microprocessor, the 8088 microprocessor, the 16-bit 80286 microprocessor, the 32-bit 386 microprocessor with 275,000 transistors, the *i*860 microprocessor containing over 1 million transistors, the 486 microprocessor, the Pentium processor containing 3 million transistors and the Pentium II and the Pentium IV with 42 million transistors.

Moore is active in philanthropy, and he has donated hundreds of millions of dollars to educational projects around the world. This includes donations to universities such as Caltech, the University of California and the University of Cambridge.

He has received many awards for his contributions to the computing field. These include the 2008 IEEE Medal of Honour which he received for pioneering technical roles in integrated circuit processing and for leadership in the semiconductor industry. The library at the Centre for Mathematical Sciences at Cambridge is named after him, as is the Moore Laboratories Building in Caltech. He received the National Medal in Technology from President George Bush in 1990.

42.1 Intel

Robert Noyce and Gordon Moore founded Intel in 1968. It is the largest semiconductor manufacturer in the world, with major plants in the United States, Europe and Asia. It has played an important role in shaping the computing field since its invention of the microprocessor in 1971. It has subsequently introduced a suite of innovative microprocessors.

The initial focus of Intel was on semiconductor memory products and to create large-scale integrated (LSI) semiconductor memory. The company made a major impact on the computer industry with its introduction of the Intel 4004 microprocessor in 1971. This was the world's first microprocessor, and although it was initially developed as an enhancement to allow users to add more memory to their units, it soon became clear that the microprocessor had great potential for everything from calculators to cash registers and traffic lights (Fig. 42.2).

The microprocessor is essentially a computer on a chip, and its invention made handheld calculators and personal computers (PCs) possible. Intel's microprocessors are used on the majority of personal computers and laptops around the world.

The 4004 was designed by Ted Hoff, and it contained a central processing unit (CPU) on one chip. It contained 2,300 transistors on a one-eighth-by-one-sixth-inch chip. The power of the large ENIAC developed (discussed in an earlier chapter), which used 18,000 vacuum tubes and took up the space of an entire room, was now available on a small chip.

Intel has introduced a suite of more and more powerful microprocessors since the 4004. These include the 8-bit 8080 microprocessor which was introduced in 1974. This was the first general-purpose microprocessor, and it sold for $360: i.e. a whole computer on one chip was sold for $360, while conventional computers sold for thousands of dollars. The 8080 soon became the industry standard, and Intel became the industry leader in the 8-bit market.

Fig. 42.2 Intel 4004 microprocessor

The 16-bit 8086 was introduced in 1978, and it was later used by IBM for its personal computer.[1] It released the 80486 microprocessor in 1989, and this was described by Business Week as *a verifiable mainframe on a chip*. It had 1.2 million transistors and the first built-in math coprocessor. It released the Pentium generation of microprocessor in 1993.

Intel dominates the microprocessor market and has a broad product line including motherboards, flash memory, switches and routers. It has made a major contribution to the computing field.

42.2   Moore's Law

Moore observed that over a period of time (from 1958 up to 1965) that the number of transistors on an integrated circuit doubled approximately every year. This led him to formulate what became known as *Moore's law* in 1965 [Mor:65], which predicted that this trend would continue for at least another 10 years. He refined the law in 1975 and predicted that a doubling in transistor density would occur every 2 years for the following 10 years.

His prediction of *exponential growth* in transistor density has proved to be accurate over the last 50 years, and the capabilities of many digital electronic devices are linked to Moore's law.

The exponential growth in the capability of processor speed, memory capacity and so on is all related to this law. It is likely that the growth in transistor density will slow to a doubling of density every 3 years by 2015.

The phenomenal growth in productivity is due to continuous innovation and improvement in manufacturing processes. It has led to more and more powerful computers running more and more sophisticated applications.

---

[1] IBM initially used the Intel 8088 microprocessor for the PC.

Chapter 43
Grace Murray Hopper

Rear Admiral Grace Murray Hopper was an American mathematician, computer scientist, computer pioneer and an American Navy naval officer. She was one of the earliest computer programmers, and she programmed the Harvard Mark I computer and its successors. She played an important role in the development of programming languages and compilers, programming language constructs, data processing and the COBOL programming language (Fig. 43.1).

She was born in New York in 1906, and she studied mathematics and physics at Vassar College. She graduated with a distinguished degree in 1928 and obtained a Master's degree in mathematics from Yale in 1930. She earned her Ph.D. in mathematics from Yale in 1934. She began teaching mathematics at Vassar College in 1931 and became an associate professor in 1941. She married Vincent Foster Hopper in 1930, but he died during the Second World War. They did not have children, and she did not remarry.

She joined the US Navy when the United States entered the Second World War, and this gave her the opportunity to contribute to the development of early computing machines. She was assigned to the Bureau of Ordnance Computation Project at Cuft Laboratories at Harvard and became familiar with the Harvard Mark I computer. This calculating machine had been designed by Howard Aiken and built by IBM. She became the third person to program the Mark I, and she went on to work on the Mark II and Mark III computers. She joined the staff of Harvard as a research fellow in engineering sciences and applied physics at the Computation Laboratory at the end of the Second World War. She coined the term *computer bug* when she traced an error in the Mark II computer to a moth stuck in one of the relays. The bug was carefully removed and taped to a daily logbook, and the term is now ubiquitous in the computer field.

She joined the Eckert-Mauchly Computer Corporation as a Senior Mathematician in 1949. The company was taken over by Remington Rand in 1950 and merged with the Sperry Corporation in 1955. She worked on the UNIVAC computer and developed the concept of a compiler during this period. She recognized the need for a friendly programming language, as programming in binary machine code was tedious and error prone. She believed that the development of a user-friendly

G. O'Regan, *Giants of Computing: A Compendium of Select, Pivotal Pioneers*,
DOI 10.1007/978-1-4471-5340-5_43, © Springer-Verlag London 2013

Fig. 43.1 Grace Murray Hopper and UNIVAC (Courtesy of Smithsonian Institute)

language would encourage wider use of computers and recognized that libraries of code could help to reduce errors and duplication of effort. This led to her idea for a compiler that would act as an intermediate program that would translate the program instructions into machine code that could be understood by the computer. This would allow programmers to employ the friendly and intuitive notation of a high-level programming language, instead of writing lengthy instructions in binary code. Programming in machine code is time consuming and error prone, as binary code consists of strings of 0s and 1s, and mistakes are easy to make and time consuming to identify. Hopper's solution to this problem was to write a compiler program that freed programmers from having to write repetitive binary code.

Her first compiler, the A-O, appeared in 1949, and it used symbolic mathematical code to represent binary code combinations. She followed this with the B-0, or *Flow-Matic* compiler in 1952, and this is considered the first English language data processing compiler. She contributed to standardizing compilers and compiler verification.

She recognized the need for a user-friendly programming language that would be easy for business users. She was closely involved in the development of COBOL, which was the first business-oriented programming language. She was the technical adviser to the CODASYL committee that defined the specification of the language, and the Flow-Matic compiler was employed to assist in its development. She designed manuals and tools for the language. COBOL was introduced in 1959, and Hopper participated in public demonstrations of the first COBOL compiler. The language provided a degree of machine independence, as the source was written once, and then compiled into the machine language of the targeted machine.

She retired from Sperry in 1971 and she retired from the Navy in 1986. She was promoted to Admiral Hopper in 1985, and she retired at the rank of Rear Admiral. She became a senior consultant to Digital Equipment Corporation in 1986, and remained there for several years.

She has received various awards in recognition of her contributions to the computing field. She received the Naval Ordnance Development Award for her pioneering application programming work on the Mark I, II and III computers. She became a Distinguished Fellow of the British Computer Society in 1973. She received the Defense Distinguished Service Medal from the Department of Defense in 1986 on her retirement from the Navy. She received the National Medal in Technology from President George Bush in 1991. She has received several honorary degrees. She died in 1992.

43.1 COBOL

The Common Business-Oriented Language (COBOL) was the first business programming language, and it was introduced in 1959. It was developed by Grace Murray Hopper and a committee of computer professionals called the Conference on Data Systems Languages (CODASYL). The committee included representatives from industry, universities in the United States and the US government.

The goal was to develop a user-friendly language that would be easy to program and to improve the readability of software source code for business users. The language has an English-like syntax that is designed to make it easy for the average business user to learn the language. It is designed for business use and is not suitable for scientific applications.

There are only three data types in the language, and these are numeric, alphanumeric and alphabetic. These data types deal with numbers and strings of text. The language allows for these to be grouped into arrays and records, so that data may be tracked and organized better. The notation is very verbose, for example, the division of two numbers is given by the following statement:

DIVIDE A BY B GIVING C REMAINDER D

COBOL was the first computer language whose use was mandated by the US Department of Defense. A typical COBOL application is often very large and may be over a million lines of code. Many of these applications are quite old and remain in use today. The language remains popular in business today, and there is an object-oriented version of the language.

The Gartner Group estimated that of the 300 billion existing lines of code in 1997, 240 billion (or 80 %) were written in COBOL. One key advantage of COBOL is that it is very easy to understand a program written in language, and that there is a minimal need for documentation.

Chapter 44
John von Neumann

John von Neumann was a Hungarian/American mathematician who made fundamental contributions to mathematics, physics, set theory, computer science, economics and quantum mechanics (Fig. 44.1).

He was born as Neumann, János Lajos in Budapest, Hungary, in 1903, and his father received the aristocratic title *von* for his contributions to the Austrian-Hungarian Empire. He earned his Ph.D. in mathematics at the age of 22 from Pázmány Péter University in Budapest in 1925.

His Ph.D. degree was concerned with the axiomatization of set theory and dealt with the problem of *Russell's paradox*. Bertrand Russell had posed the question whether the set S of all sets that do not contain themselves as members contains itself as a member, i.e. does $S \in S$? In either case, a contradiction arises since if $S \in S$, then as S is a set that does not contain itself as a member and therefore $S \notin S$. Similarly, if $S \notin S$, then S is a set that does not contain itself as a member, and therefore $S \in S$.

von Neumann showed how the contradiction in set theory can be avoided in two ways. One approach is to employ the axiom of foundation, and the other approach is to employ the concept of a class.

He taught at the University of Berlin from 1926 to 1930 and moved to the United States in 1930 to take a position at the newly established Institute for Advanced Studies at Princeton, New Jersey. He was to remain a mathematics professor at the Institute for the remainder of his life.

He considered the problem of the axiomatization of quantum theory and showed how the physics of quantum theory could be reduced to the mathematics of linear operators on Hilbert spaces [VN:32]. His theory included the Schrödinger wave mechanical formulation and Heisenberg matrix mechanical formulation as special cases. von Neumann and Birkhoff later proved that quantum mechanics requires a logic that is quite different from classical logic.

He made contributions to economics and game theory. He was interested in practical problems and worked as a consultant to the US Navy, IBM, the CIA and the Rand Corporation. He became an expert in the field of explosions and discovered

G. O'Regan, *Giants of Computing: A Compendium of Select, Pivotal Pioneers*,
DOI 10.1007/978-1-4471-5340-5_44, © Springer-Verlag London 2013

that large bombs are more devastating if they explode before touching the ground.
He contributed to the development of the hydrogen bomb and to improving methods
to utilize nuclear energy.

He gave his name to the *von Neumann architecture* used in almost all computers.
Eckert and Mauchly were working with him on this concept during their work
on ENIAC and EDVAC, but their names were removed from the final report
due to their resignation from the University of Pennsylvania to form their own
computer company. von Neumann architecture includes a central processing unit
which includes the control unit and the arithmetic unit, an input and output unit and
memory.

He also created the field of cellular automata and is the inventor of the *merge
sort algorithm* (in which the first and second halves of an array are each sorted
recursively and then merged). He also invented the *Monte Carlo* method that allows
complicated problems to be approximated through the use of random numbers.

He died at the relatively young age of 54 in 1957. There is an annual von
Neumann medal awarded by the IEEE for outstanding achievements in computer
science. The next section discussed von Neumann architecture.

44.1   von Neumann Architecture

The earliest computers were fixed program machines and were designed to do a
specific task. This proved to be a major limitation as it meant that a complex manual
rewiring process was required to enable the machine to solve a different problem.

The computers used today are general-purpose machines designed to allow
a variety of programs to be run on the machine. The fundamental architecture
underlying modern computers was described by von Neumann and others in the
mid-1940s. It is known as the von Neumann architecture.

Fig. 44.2 von Neumann
architecture

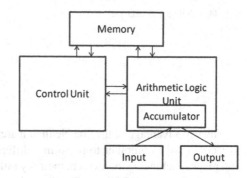

Table 44.1 von Neumann architecture

Component	Description
Arithmetic unit	The arithmetic unit is capable of performing basic arithmetic operations
Control unit	The program counter contains the address of the next instruction to be executed. This instruction is fetched from memory and executed. This is the basic fetch-and-execute cycle
	The control unit contains a built-in set of machine instructions
Input–output unit	The input and output unit allows the computer to interact with the outside world
Memory	There is a one-dimensional memory that stores all of the program instructions and data
	Program instructions and data are usually kept in different areas of memory
	The memory may be written to or read from; i.e. it is random access memory (RAM)
	The program instructions are binary values, and the control unit decodes the binary value to determine the particular instruction to execute

It arose on work done by von Neumann, Eckert, Mauchly and others on the design of the EDVAC. This was the successor to ENIAC, and von Neumann's draft report on EDVAC [VN:45] described the new architecture. The EDVAC was built in 1949 (Fig. 44.2).

von Neumann architecture led to the birth of stored program computers where a single store is used for both machine instructions and data. Its key components are shown in Table 44.1.

The key approach to building a general-purpose device according to von Neumann was in its ability to store not only its data and intermediate results of computation but also to store the instructions or commands for the computation. The computer instructions can be part of the hardware for specialized machines, but for general-purpose machines, the computer instructions must be as changeable as the data that is acted upon by the instructions. His insight was to recognize that both the machine instructions and data could be stored in the same memory (Fig. 44.3).

Fig. 44.3 Fetch-and-execute
cycle

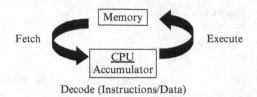

Decode (Instructions/Data)

The key advantage of the von Neumann architecture was that it was much simpler
to reconfigure a computer to perform a different task. All that was required was to
enter new machine instructions in memory rather than physically rewiring a machine
as was required with ENIAC. The limitations of von Neumann architecture include
that it is limited to sequential processing rather than parallel processing.

Chapter 45
Ken Olsen

Kenneth Harry Olsen was an American engineer and entrepreneur who co-founded Digital Equipment Corporation (DEC) with Harlan Anderson. DEC dominated the minicomputer era from the 1960s to 1980s, with its PDP and VAX series of computers (Fig. 45.1).

Olsen was born in Connecticut in 1926 and became interested in electronics, gadgets and radios at an early age. He served in the US Navy between 1944 and 1946 and was involved in a project to build a flight simulator at the Office of Naval Research. He studied electrical engineering at the Massachusetts Institute of Technology after the war, and he received a bachelor's degree in 1950 and a master's degree in 1952. He worked at MIT's Lincoln computer laboratory after graduation and was part of the air defence team that was working on the SAGE Air Defence System. He was involved in building a transistorized research computer and also worked on ways to improve the magnetic core memory developed by his MIT colleague, Jay Forrester.

Olsen and an MIT colleague, Harlan Anderson, founded Digital Equipment Corporation (DEC) in 1957, with venture capital from American Research and Development Corporation. It was a forward-thinking innovative company that would become (for a period of time) the second largest computer company in the world. DEC introduced the PDP family of minicomputers in the 1960s and the 32-bit VAX series of minicomputers in the 1970s. Olsen retired from the company in 1992.

He received various awards for his contributions to the computing field and was named by *Fortune* magazine as America's most successful entrepreneur in 1986. He received the IEEE Engineering Leadership Recognition Award in the same year. He received the US National Medal of Technology in 1993. He was a co-founder of the Computer History Museum with Gordon Bell (discussed in an earlier chapter) and was a fellow of the museum.

G. O'Regan, *Giants of Computing: A Compendium of Select, Pivotal Pioneers*, 209
DOI 10.1007/978-1-4471-5340-5_45, © Springer-Verlag London 2013

Fig. 45.1 Ken Olsen

45.1   Digital Equipment Corporation

DEC was founded in 1957 by Ken Olsen and Harlan Anderson who were engineers working at MIT's Lincoln Laboratory. It became a leading vendor of computer systems from the 1960s to the 1990s, and its PDP and VAX computers were very popular with the engineering and scientific communities. It became the second largest computer company in the world at its peak in the late 1980s, when it employed over 120,000 people and had annual revenues of $14 billion.

Its first computer, the *Programmed Data Processor* (PDP-1), was released in 1961. It was a relatively inexpensive computer and cost $110,000. The existing IBM mainframes cost over $2 million, and so DEC's minicomputers were relatively affordable to businesses. It was a simple and reasonably easy to use computer with 4,000 words of memory. It was an 18-bit machine and one of the earliest computer games, *Spacewar*, was developed for this machine. The PDP series of minicomputers were elegant and reasonably priced and dominated the new minicomputer market segment. They were an alternative to the multimillion dollar mainframes offered by IBM to large corporate customers. Research laboratories, engineering companies and other professions requiring large computer all used DEC's minicomputers (Fig. 45.2).

Olsen and Anderson were engineers rather than managers, and DEC's culture was that of an engineering company. The company was divided into competing product groups, with line managers given complete authority to get their jobs done. The only control was budgetary, and the groups were given complete freedom in product development. There was a certain lack of central direction, but it meant that each group was, in effect, in charge of its own destiny. If things were not working effectively, then the budget would dry up, and staff would be reassigned to other product groups.

The PDP-8 minicomputer was a 12-bit machine with a small instruction set, and it was released in 1965. It was a major commercial success for DEC with many sold

Fig. 45.2 PDP-1 computer

to schools and universities. The PDP-11 was a highly successful series of 16-bit minicomputer, and it remained a popular product for over 20 years from the 1970s to the 1990s.

Its next generation of computers was the 32-bit VAX series of computers, and these were introduced following the return of Gordon Bell as the VP of engineering in 1972. The VAX series were derived from the PDP-11, and it was the first widely used 32-bit minicomputer. The VAX 11/780 was released in 1978, and it was a major success for the company. The VAX product line was a competitor to the IBM System/370 series of computers. It used the Virtual Operating System known as VMS.

The rise of the microprocessor and microcomputer led to the availability of low-cost personal computers, and this later challenged DEC's product line. DEC was slow in recognizing the importance of these developments, and Olsen's statement from the mid-1970s *There is no need for any individual to have a computer in his home* suggests that DEC were totally unprepared for this new threat to their business.

DEC responded with its own personal computer after the launch of the IBM PC. The DEC machine easily outperformed the PC, but it was more expensive and incompatible with the IBM PC hardware and software. DEC's microcomputer efforts were a failure, but its PDP and VAX products were continuing to sell. By the late 1980s, DEC was threatening IBM's number one spot in the industry. However, the increasing power of the newer generations of microprocessors began to challenge DEC's minicomputer product range.

Ultimately, Olsen and the company were too late in responding to the paradigm shift in the industry, and this proved to be fatal for the company. Digital was a strong engineering company, and it seemed to believe that the engineering excellence of its computer products would drive the market and be sufficient for its financial success. It seemed to believe that it knew better than its customers on what was needed for the industry, and these views bordered on technological arrogance. Olsen was sceptical of personal computers and thought of them as merely toys for playing video games, rather than as serious machines in their own right. He seemed to believe that a customer would only be interested in a computer for serious scientific work, such as the elegant machines developed by DEC.

He failed to recognize the importance of the release of the Apple computer and its successors and the IBM personal computer. The company failed to adapt in time to the personal computer market, and its sales declined from the early 1990s. Further, indecision and infighting inside the company delayed an appropriate response to the challenges.

Olsen retired in 1992 and Robert Palmer became the new CEO. He was given the responsibility to return the company to profitability, and he attempted to change the business culture and to sell off noncore businesses. This led to massive layoffs, and eventually, Compaq acquired Digital in 1998 for $9.8 billion. Compaq later merged with HP.

Chapter 46
David Parnas

David Lorge Parnas has been influential in the computing field, and his ideas on the specification, design, implementation, maintenance and documentation of computer software remain relevant today. He has won numerous awards (including ACM best paper award in 1979, two most influential paper awards from ICSE in 1978 and 1984, the ACM SigSoft outstanding researcher award in 1998 and an honorary doctorate from the ETH in Zurich and the Catholic University of Louvain in Belgium) for his contribution to computer science. Software engineers today continue to use his ideas in their work (Fig. 46.1).

He was born in the United States on February 10, 1941, and he studied at Carnegie Institute of Technology from 1957 to 1965. He was awarded a bachelor's degree in electrical engineering in 1961, and this was followed by a Master's degree, and he earned a Ph.D. degree from Carnegie in 1965. He has worked in both industry and academia, and his approach aims to achieve a middle way between theory and practice. His research has focused on real industrial problems that engineers face and on finding solutions to these practical problems. Several organizations such as Phillips in the Netherlands, the Naval Research Laboratory (NRL) in Washington, IBM Federal Systems Division and the Atomic Energy Board of Canada have benefited from his expertise.

His ideas on the specification, design, implementation, maintenance and documentation of computer software remain important. He advocates a solid engineering approach to the development of high-quality software and argues that the role of the engineer is to apply scientific principles and mathematics to design and develop useful products. He argues that computer scientists should be educated as engineers and provided with the right scientific and mathematical background to do their work effectively.

Parnas has made a strong contribution to software engineering, and he has written over 200 research papers. These include contributions to requirements specification, software design, software inspections, testing, tabular expressions, predicate logic, precise mathematical documentation and ethics for software engineers. He has

Fig. 46.1 David Parnas
(Courtesy of Hubert
Baumeister)

made solid contributions to industry and teaching. His reflections on software
engineering are valuable and contain insight and wisdom gained over a long career.
His contributions to software engineering include (Table 46.1):

He played an important role in criticizing the Strategic Defence Initiative (SDI)
launched by President Reagan in the early 1980s. The *star wars* initiative planned
to use ground- and space-based systems to protect the United States from nuclear
attack. It threatened to reignite a new arms race, and Parnas criticized the SDI
on the grounds that it was too ambitious and unrealistic. Next, we discuss tabular
expressions in detail.

46.1   Tabular Expressions

Tables of constants have been used for millennia to define mathematical functions.
The tables allow the data to be presented in an organized form that is easy
to reference and use. The data in a table provides an explicit definition of the
mathematical function and allows the computation of the function for a particular
value to be easily done. Tables are used in schools (e.g. where primary school
children are taught multiplication tables and high school students refer to sine or
cosine tables).[1] The invention of electronic calculators may lead to a reduction in
the use of tables as students may compute the values of functions immediately.

Tabular expressions are a generalization of tables, in which constants can be
replaced by more general mathematical expressions. Conventional mathematical
expressions are a special case of tabular expressions. In fact, everything that can be
expressed as a tabular expression can be represented by a conventional expression.
Tabular expressions can represent sets, relations, functions and predicates and

---

[1] In today's world of computers and calculators, the use of tables may become a thing of the past.

Table 46.1 Parnas's achievements

Area	Description
Tabular expressions	Tabular expressions are mathematical tables that are employed for specifying the requirements of a system. They enable complex predicate logic expressions to be represented in a simpler form
Precise mathematical documentation	He advocates the use of mathematical documents that are precise and complete. These include the system requirements, system design, software requirements, module interface specification, and module internal design
Requirements specification	His approach to requirements specification (developed with Kathryn Heninger and others) is mathematical. It involves the use of mathematical relations to specify the requirements precisely
Software design	He introduced the revolutionary *information-hiding* principle, which allows software to be *designed in a way to deal with change*. A module is characterized by its knowledge of a design decision (*secret*) that it hides from all other modules. Every information-hiding module has an *interface* that provides the only means to access the services provided by the modules. *The interface hides the module's implementation.* Information hiding is used in object-oriented programming
Predicate logic	He introduced an approach to deal with *undefined values*[a] in predicate logic expressions, which preserves a two-valued logic
Software inspections	His approach to software inspections is quite distinct from the Fagan inspection methodology. The reviewers are required to carry out mathematical analysis to provide answers to questions posed by the author. This involves the production of mathematical tables
Teaching	He has taught at various universities including McMaster University in Canada. His ideas on an engineering education for software engineers are important
Industry contributions	He played a key role in defining the requirements of the A7 aircraft and the formal inspection of safety critical software for the nuclear power plant at Darlington
Ethics for software engineers	He has argued that software engineers have a professional responsibility as engineers. They need to build safe products and to accept responsibility for their design decisions

[a]His approach allows undefinedness to be addressed in predicate calculus while maintaining the 2-valued logic. A primitive predicate logic expression that contains an undefined term is considered false in the calculus. This is an unusual way of dealing with undefinedness

conventional expressions. A tabular expression may also be represented by a conventional expression, but its advantage is that it is easier to read and use than a complex conventional expression. A tabular expression replaces a complex expression with a set of simpler expressions, and they have been applied to practical problems such as the precise documentation of the system requirements of the A7 aircraft [Hen:80] (Table 46.2).

A tabular expression provides a useful way to define a piecewise continuous function, and it is relatively easy to demonstrate that all cases have been considered

Table 46.2 Applications
of tabular expressions

Specify requirements
Specify module interface design
Description of implementation of module
Mathematical software inspections

		$y = 5$	$y > 5$	$y < 5$	H_2
H_1	$x \geq 0$	0	y^2	$-y^2$	G
	$x < 0$	x	$x+y$	$x-y$	

Fig. 46.2 Tabular expressions (normal table)

in the definition. The standard definition of a piecewise-defined function may miss a case or to give an inconsistent definition. The evaluation of a tabular expression is easy once the type of the tabular expression is known.

Parnas and others have identified a collection of tabular expressions that may be employed to document the system requirements. He gave a precise meaning to each type of tabular expressions in terms of their component expressions [Par:92], and a more general model was proposed by Janicki [Jan:97]. Parnas and others have proposed a general mathematical foundation for tabular expressions.

The function $f(x,y)$ is defined in the tabular expression below. The tabular expression consists of headers and a main grid. The headers define the domain of the function, and the main grid gives the definition. It is easy to see that the function is defined for all values on its domain as the headers are complete. It is also easy to see that the definition is consistent as the headers partition the domain of the function (Fig. 46.2).

The evaluation of the function for a particular value (x,y) involves determining the appropriate row and column from the headers of the table and computing the grid element for that row and column.

For example, the evaluation of $f(2,3)$ involves the selection of row 1 of the grid (as $x = 2 \geq 0$ in H_1) and the selection of column 3 (as $y = 3 < 5$ in H_2). Hence, the value of $f(2,3)$ is given by the expression in row 1 and column 3 of the grid, i.e. $-y^2$ evaluated with $y = 3$ resulting in -9. The table simplifies the definition of the function.

The more general definition of tabular expressions allows for multidimensional tables including multiple headers and supports rectangular and non-rectangular tables. Usually, the headers contain predicate expressions, whereas the grid usually contains terms. However, the role of the grid and the headers changes depending on the type of table being considered.

There are many types of tabular expressions such as the normal function table, the inverted function table, the vector table and the mixed function table. The reader is referred to [ORg:06].

46.2 Classical Engineering Education

The construction of bridges was problematic in the nineteenth century, and many people who presented themselves as qualified to design and construct bridges did not have the required knowledge and expertise. Consequently, many bridges collapsed, endangering the lives of the public. This led to legislation requiring an engineer to be licensed by the Professional Engineering Association prior to practising as an engineer. These engineering associations identify a core body of knowledge that the engineer is required to possess, and the licensing body verifies that the engineer has the required qualifications and experience. The licensing of engineers by most branches of engineering ensures that only personnel competent to design and build products actually do so. This in turn leads to products that are safe for the public to use. In other words, the engineer has a responsibility to ensure that the products are properly built and are safe to use.

Parnas argues that traditional engineering be contrasted with the software engineering discipline where there is no licensing mechanism and where individuals with no qualifications can participate in the design and building of software products.[2]

Classical engineering requires the engineer to state precisely the requirements that the software product is to satisfy and then to produce designs that will meet these requirements. Engineers provide a precise description of the problem to be solved; they then proceed to producing a design and validate the correctness of the design; finally, the design is implemented and testing is performed to verify its correctness with respect to the requirements.

Classical engineers employ mathematics to analyze their design to determine its correctness. The level of mathematics employed will depend on the particular application. The term *engineer* is generally applied only to people who have attained the necessary education and competence to be called engineers and who base their practice on mathematical and scientific principles. Often in computer science, the term engineer is employed rather loosely to refer to anyone who builds things, rather than to an individual with a core set of knowledge, experience and competence.

Programmers are like engineers in the sense that they design and build products. Parnas argues that they need an appropriate education to design and develop software and to apply scientific and mathematical principles to their work. He is a strong advocate of the classical engineering approach, and he argues that computer scientists should be taught mathematics and design to enable them to build high-quality and safe products.

He has argued that computer science courses tend to include a small amount of mathematics, whereas mathematics is a significant part of an engineering course. Parnas argues that students are generally taught programming and syntax, but not

---

[2]Modern HR recruitment specifies the requirements for a particular role, and interviews with candidates aim to establish that the candidate has the right education and experience for the role.

how to design and analyze software. He advocates a solid engineering approach to the teaching of mathematics with an emphasis on its application to developing and analyzing product designs.

He argues that software engineers need education on engineering mathematics, specification and design, converting designs into programs, software inspections and testing. The education should enable the software engineer to produce well-designed programs that will correctly implement the requirements. He has stated that the term "engineer" is a title that is awarded on merit, but it also places responsibilities on its holder. Professional engineers are required to follow rules of good practice and to object when the rules are violated.

Chapter 47
Dennis Ritchie

Dennis MacAlistair Ritchie was an American computer scientist who is famous for developing the C programming language at Bell Labs. He also codeveloped the UNIX operating system with Ken Thompson (Fig. 47.1).

He was born in New York in 1941, and he did a degree at Harvard University in 1963. He earned a Ph.D. in physics and applied mathematics from the university in 1967.

He joined Bell Labs in 1967 and was involved in the Multics operating system project. He designed and implemented the C programming language at Bell Labs in the early 1970s. The origin of this language is closely linked to the development of the UNIX operating system, and C was originally used for systems programming. It later became very popular for both systems and application programming and influenced the development of other languages such as C++ and Java.

Brian Kernighan wrote the first tutorial on C, and Kernighan and Ritchie later wrote the popular book *The C Programming Language* [KeR:78]. Ritchie later became head of Lucent Technologies Systems Software Research Department.

He jointly received the ACM Turing Award with Ken Thompson in 1983, in recognition of their achievements in the implementation of the UNIX operating system. They received the IEEE Pioneer Award in 1994 and the National Medal in Technology from Bill Clinton in 1999. Ritchie retired from Lucent in 2007 and died in 2011.

47.1 C Programming Language

Richie developed the C programming language at Bell Lab in 1972, and it became a popular programming language used widely in the industry. It is a general-purpose and systems programming language.

It was originally designed as a language to write the kernel for the UNIX operating system. It had been traditional up to then to write the operating system kernel in assembly languages, and the use of a high-level language such as C was

G. O'Regan, *Giants of Computing: A Compendium of Select, Pivotal Pioneers*,
DOI 10.1007/978-1-4471-5340-5_47, © Springer-Verlag London 2013

Fig. 47.1 Ken Thompson and Dennis Ritchie with President Clinton in 1999

a paradigm shift. The successful use of C to write the UNIX kernel led to its use as a systems programming language on several other operating systems (e.g. Windows and Linux). C also influenced later language development. The language is described in detail in [KeR:78].

The language provides high-level and low-level capabilities, and a C program that is written in ANSI C may be compiled for a very wide variety of computer platforms and operating systems (with minimal changes to the source code).

C is a procedural programming language and includes conditional statements such as the *if statement*, the *switch statement*, iterative statements such as the *while* statement and *do* statement and the assignment statement.

- If statement

```
if  (A==B)¹
      A = A + 1;
else
      A = A − 1²;
```

- Assignment statement

```
i = i + 1;
```

---

[1] One common error in C programs is writing "=" instead of "==". This totally alters the meaning of the statement.

[2] The semicolon in Pascal is used as a statement separator, whereas it is used as a statement terminator in C.

One of the first programs that people write in C is the *Hello world* program. This is given by

```
main()
{
    printf("Hello, World\n");
}
```

C includes several predefined data types including integers and floating point numbers.

int	(integer)
long	(long integer)
float	(floating point real)
double	(double precision real)

It allows more complex data types to be created using *structs* (these are similar to records in Pascal). It allows the use of pointers to access memory locations, and this allows the memory locations to be directly referenced and modified. The result of the following example is to assign 5 to the variable *x*, and the memory address of variable *x* is given by &*x*.

```
int x;
int *ptr_x;

x = 4;
ptr_x = &x;
*ptr_x = 5;
```

C is a block-structured language, and a program is structured into functions (or blocks). Each function block contains its own variables and functions. A function may call itself (i.e. *recursion is allowed*).

```
if (a==b)
    a++;                 ....Program fragment A
else
    a--

if (a = b)
    a++;                 ....Program fragment B
else
    a--
```

One key criticism of C is that it is easy to make errors in C programs, and to thereby produce undesirable results. For example, one of the easiest mistakes to make is to accidently write the assignment operator (=) for the equality operator (==). This totally changes the meaning of the original statement as can be seen in program fragments A and B above.

Both program fragments are syntactically correct, and the intended meaning of a program has been changed by writing "=" instead of "==". The philosophy of C is to allow statements to be written as concisely as possible, and this is potentially dangerous.[3] The use of pointers potentially leads to problems as uninitialized pointers may point anywhere in memory, and the program may potentially overwrite anywhere in memory. Therefore, the effective use of C requires experienced programmers, well-documented source code and formal peer reviews of the source code by other team members.

47.2   UNIX

The UNIX operating system was developed by Ken Thompson, Dennis Ritchie and others at Bell Labs in the early 1970s. It is a multitasking and multi-user operating system written almost entirely in C. UNIX arose out of work by Massachusetts Institute of Technology, General Electric and Bell Labs on the development of a general time-sharing operating system called *Multics*.

Bell Labs decided in 1969 to withdraw from the Multics project and to use General Electric's GECOS operating system. However, several of the Bell Lab researchers decided to continue the work on a smaller-scale operating system using a Digital PDP-7, and later they used a PDP-11 computer. The result of their work was UNIX, and it became a widely used operating system. It was used initially by universities and the US government but later became popular in industry.

It is a powerful and flexible operating system and is used on a variety of machines from micros to supercomputers. It is designed to allow several programmers to access the computer at the same time and to share its resources, and it offers powerful real-time sharing of resources.

It includes features such as *multitasking* which allows the computer to do several things at once, *multi-user* capability which allows several users to use the computer at the same time, *portability* of the operating system which allows it to be used on several computer platforms with minimal changes to the code and a collection of tools and applications. There are three levels of the UNIX system: *kernel*, *shell* and *tools and applications*.

---

[3]It is easy to write a one-line C program that is incomprehensible. The maintenance of poorly written code is a challenge unless programmers follow good programming practice. This discipline needs to be enforced by formal reviews of the source code.

Chapter 48
Dana Scott

Dana Scott has made important contributions to theoretical computer science, including automata theory, and the theory of programming language semantics. He has also contributed to modal logic, topology and category theory (Fig. 48.1).

He was born in 1932 and studied mathematics at the University of California, Berkeley. He obtained his bachelor's degree in 1954 from the university and earned his Ph.D. in mathematics in 1958 from Princeton University. His doctoral advisor was the logician Alonzo Church.

He co-authored an influential joint paper entitled *Finite Automaton and Their Decision Problem* with Michael O. Rabin in 1959. This paper introduced the idea of non-deterministic finite-state machines to automata theory. They were later to receive the ACM Turing Award for the introduction of this fundamental concept, and this classic paper influenced subsequent work in the field.

He became assistant professor of mathematics at the University of California, Berkeley, and began working in modal logic. He moved to Oxford University in England in 1972 and worked with the late Christopher Strachey on mathematical foundations for the semantics of programming languages. Their work constitutes the Scott-Strachey approach to denotational semantics and has been highly influential in the programming semantics field. Scott formulated domain theory which allows programs, including recursive functions, to be given a precise mathematical meaning.

He returned to the United States in 1981 and became a professor of computer science at Carnegie Mellon University. He retired in 2003 and lives in Berkeley, California.

He has received many honours in recognition of his contributions to the theoretical computing field. He jointly received the ACM Turing Award with Michael O. Rabin in 1976. He received the EATCS Award in 2007 in recognition of his contribution to theoretical computer science.

G. O'Regan, *Giants of Computing: A Compendium of Select, Pivotal Pioneers*,
DOI 10.1007/978-1-4471-5340-5_48, © Springer-Verlag London 2013

Fig. 48.1 Dana Scott

48.1 Automata Theory

A finite-state machine (FSM) is an abstract mathematical machine that consists of a finite number of states. It includes a start state q_0 in which the machine is in initially a finite set of states Q, an input alphabet Σ, a state transition function δ and a set of final accepting states F (where $F \subseteq, Q$).

The state transition function takes the current state and an input and returns the next state. That is, the transition function is of the form

$$\delta : Q \times \Sigma \rightarrow Q$$

The transition function provides rules that define the action of the machine for each input, and it may be extended to provide output as well as a state transition. State diagrams are used to represent finite-state machines, and each state accepts a finite number of inputs. A finite-state machine may be *deterministic* or *non-deterministic*, and a deterministic machine changes to exactly one state for each input transition, whereas a non-deterministic machine may have a choice of states to move to for a particular input (Fig. 48.2).

A *non-deterministic* automaton (NFA) or non-deterministic finite-state machine is a finite-state machine where from each state of the machine and any given input, the machine may jump to several possible next states. For a deterministic finite-state automaton, the next state is uniquely determined for any input, whereas for a non-deterministic automaton, there is a choice of states. However, a non-deterministic automaton is equivalent to a deterministic automaton, in that they both recognize the same formal language (i.e. regular languages as defined in Chomsky's classification). For any non-deterministic automaton, it is possible to construct the equivalent deterministic automaton using power set construction (Fig. 48.3).

Fig. 48.2 Deterministic
finite-state machine

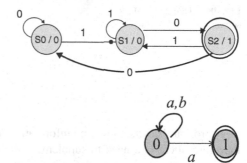

Fig. 48.3 Non-deterministic
finite-state machine

NFAs were introduced by Scott and Rabin in 1959, and an NFA is defined formally as a 5-tuple $(Q, \Sigma, \delta, q_0, F)$ as in the definition of a deterministic automaton, and the only difference is in the transition function δ.

$$\delta : Q \times \Sigma \to \mathbb{P}Q$$

Finite-state automata can compute only very primitive functions and are not an adequate model for computing. There are more powerful automata such as the *Turing machine* that is essentially a finite automaton with an infinite storage (memory). Anything that is computable is computable by a Turing machine.

48.2 Scott-Strachey Approach

The formal semantics of a programming language is concerned with defining the actual meaning of a language. Language semantics is deeper than syntax, and the theory of the syntax of programming languages is well established. A programmer writes a program according to the rules of the language. The compiler first checks the program for syntactic correctness; i.e. it determines whether the program as written is valid according to the rules of the grammar of the language. If the program is syntactically correct, then the compiler determines the meaning of what has been written and generates the corresponding machine code.[1]

The compiler must preserve the semantics of the language; i.e. the semantics are not defined by the compiler, but rather the function of the compiler is to preserve the semantics of the language. Therefore, there is a need to have an unambiguous definition of the meaning of the language independently of the compiler, and the meaning is then preserved by the compiler.

---

[1]Of course, what the programmer has written may not be what the programmer had intended.

Fig. 48.4 Denotational
semantics

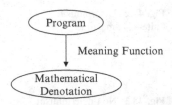

A program's syntax gives no information as to the meaning of the program, and, therefore, there is a need to supplement the syntactic description of the language with a formal unambiguous definition of its semantics.

The formal semantics of a language is given by a mathematical model that describes the possible computations described by the language. There are three main approaches, namely, *axiomatic semantics*, *operational semantics* and *denotational semantics*. These are described in more detail in [ORg:13].

Denotational semantics was developed by Christopher Strachey and Dana Scott at the Programming Research Group at Oxford, England, and it is known as the *Scott-Strachey approach* [Sto:77]. The meaning of programs is defined in terms of mathematical objects such as integers, tuples and functions, and each phrase in the language is translated into a mathematical object that is termed the *denotation* of the phrase.

The semantics of a programming language is given by a translation schema that associates a meaning (*denotation*) with each program in the language. It maps a program directly to its meaning, and it was originally called mathematical semantics, as it provides meaning to programs in terms of mathematical values such as integers, tuples and functions (Fig. 48.4).

The denotational description of a programming language is given by a set of *meaning functions* M associated with the constructs of the language. Each meaning function is of the form $M_T : T \rightarrow D_T$ where T is some construct in the language. Many of the meaning functions will be *higher order*: i.e. functions that yield functions as results. The signature of the meaning function is from syntactic domains (i.e. T) to semantic domains (i.e. D_T). A valuation map $V_T : T \rightarrow \mathbf{B}$ may be employed to check the static semantics prior to giving a meaning of the language construct.[2]

Dana Scott's contributions included the formulation of domain theory, and this allowed programs containing recursive functions and loops to be given a precise semantics.

---

[2]This is similar to what a compiler does in that if errors are found during the compilation phase, the compiler halts and displays the errors and does not continue with code generation.

48.3   Domain Theory

Domain theory is a branch of mathematics that studies *partially ordered sets*, and it has important applications to computer science. It was developed by Scott and Strachey in the early 1970s, in order to have appropriate domains in which to define the semantic functions used in denotational semantics.

The semantic domains cannot simply be sets as they need to be able to give a precise meaning to recursively defined functions. The semantics of recursion needs an explicit meaning of *nontermination* at the semantics level.

Scott introduced the bottom element ($\perp$) into each primitive semantic domain to denote and he introduced the term lifted domain $X_\perp$ to denote $X \cup \{\perp\}$, and the element $\perp$ is the least element in the domain.

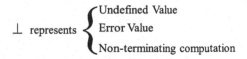

$$\perp \text{ represents } \begin{cases} \text{Undefined Value} \\ \text{Error Value} \\ \text{Non-terminating computation} \end{cases}$$

Scott used partially ordered sets to model a domain of computation. A *directed subset* of a domain has the property that any two elements of the subset have some upper bound that is an element of this subset. A *complete partial order* has the property that all directed sets have a least upper bound.

For more detailed information on domain theory and denotational semantics, see [Sch:86].

Chapter 49
John Searle

John Searle has made important contributions to the philosophy of mind, the philosophy of language and artificial intelligence (AI). He has criticized the view that artificial intelligence research will lead to conscious intelligent machines, and his *Chinese Room thought experiment* is a famous rebuttal of strong AI. He argues that computation, in itself, is not sufficient for an entity to be judged to be intelligent and conscious (Fig. 49.1).

He was born in Colorado in 1932 and was an undergraduate at the University of Wisconsin-Madison. He became a Rhodes Scholar at Oxford University in England, where he obtained his undergraduate degree in philosophy in 1955. He earned a Ph.D. in philosophy from Oxford in 1959. He held a position as lecturer in philosophy at Oxford from 1956 to 1959 and has held a position as professor of philosophy at Berkeley University since 1959. He has held several visiting professor positions around the world.

He is the author of many books on various areas of philosophy including the philosophy of language, consciousness and the philosophy of mind. He has received several awards in recognition of his contribution to philosophy, including the Jean Nicod Prize in 2000 and the National Humanities Medal in 2004.

49.1 Searle's Chinese Room

Searle's Chinese Room thought experiment is a famous paper on machine understanding. This classic paper was written by John Searle in 1980, and it presents a very compelling argument against the feasibility of the strong AI project. It rejects the claim that a machine has or will someday in the future have the same cognitive qualities as humans. Searle argues that *brains cause minds*, and that *syntax does not suffice for semantics*. There are a number of replies to Searle's position, but he has coherent arguments against these objections. He defines the terms *strong* and *weak AI* as follows [Sea:80].

Fig. 49.1 John Searle

49.1.1   Strong AI

The computer is not merely a tool in the study of the mind, rather *the appropriately programmed computer really is a mind* in the sense that computers given the right programs can be literally said to *understand* and have other cognitive states.

49.1.2   Weak AI

Computers just *simulate* thought, their seeming understanding isn't real understanding (just as-if), their seeming calculation is only as-if calculation, etc. Nevertheless, computer simulation is useful for *studying* the mind (as for studying the weather and other things).

49.2   The Thought Experiment

A man is placed into a closed room into which Chinese writing symbols are input to him. He is given a rulebook that shows him how to manipulate the symbols to produce Chinese output. He has no idea as to what each symbol means but with the rulebook he is able to produce the Chinese output. This allows him to communicate with the other person and appear to understand Chinese. The rulebook allows him to answer any question posed, without the slightest understanding of what he is doing or what the symbols mean.

1. Chinese characters are entered through slot 1.
2. The rulebook is employed to construct new Chinese characters.
3. Chinese characters are outputted to slot 2.

Fig. 49.2 Searle's Chinese
Room

你怎么样   我很好，谢谢

Input   Output

The question *Do you understand Chinese?* could potentially be asked, and the
rulebook would be consulted to produce the answer *Yes, of course*, despite of the
fact that the person inside the room has not the faintest idea of what is going on. It
will appear to the person outside the room that the person inside is knowledgeable
on Chinese, where the reality is that the person inside is just following rules without
any understanding (Fig. 49.2).

The process is essentially that of a computer program which has an input,
performs a computation based on the input and then finally produces an output.
Searle has essentially constructed a machine which can never be mental. Changing
the program essentially means changing the rulebook, and this does not increase
understanding. The strong artificial intelligence thesis states that given the right
program, *any* machine running it would be mental. However, Searle argues that the
program for this Chinese room would not understand anything, and that therefore
the strong AI thesis must be false. In other words, *Searle's Chinese room argument
is a rebuttal of strong AI* by showing that a program running on a machine that
appears to be intelligent has no understanding whatsoever of the symbols that it is
manipulating. That is, given any rulebook (i.e. program), the person would never
understand the meanings of those characters that are manipulated.

That is, just because the machine acts like it knows what is going on, it actually
only knows what it is programmed to know. It differs from humans in that it is not
aware of the situation like humans are. This suggests that machines may not have
intelligence or consciousness, and the Chinese room argument applies to any Turing
equivalent computer simulation.

There are several rebuttals of Searle's position,[1] and one well-known rebuttal
attempt is the *systems reply* argument. This rebuttal argues that if a result associated
with intelligence is produced, then intelligence must be found somewhere in the
system. The proponents of this argument draw an analogy between the human brain
and its constituents. None of its constituents have intelligence but the system as a
whole (i.e. the brain) exhibits intelligence. Similarly, the parts of the Chinese room
may lack intelligence, but the system as a whole is intelligence. Searle has provided
coherent objections to rebuttals of his position.

---

[1]I don't believe that Searle's argument proves that Strong AI is impossible, and while there are
objections to the "systems reply" rebuttal, it does seem reasonable to argue that there is some
intelligence in the system. However, I am not expecting to see intelligent machines anytime soon.

Chapter 50
Claude Shannon

Claude Shannon was an American mathematician and engineer who made fundamental contributions to computing. He was the first person[1] to see the applicability of Boolean algebra to simplify the design of circuits and telephone routing switches. He showed that Boole's symbolic logic developed in the nineteenth century provided the perfect mathematical model for switching theory and for the subsequent design of digital circuits and computers (Fig. 50.1).

He was born in Michigan in 1916, and his primary degree was in mathematics and electrical engineering at the University of Michigan in 1936. He earned a Ph.D. in mathematics from the Massachusetts Institute of Technology (MIT) in 1940.

His influential *Master's thesis is a key milestone in computing*, and it shows how to lay out circuits according to Boolean principles. It provides the theoretical foundation of switching circuits, and *his insight of using the properties of electrical switches to do Boolean logic is the basic concept that underlies all electronic digital computers*.

Shannon realized that you could combine switches in circuits in such a manner as to carry out symbolic logic operations. This allowed binary arithmetic and more complex mathematical operations to be performed by relay circuits. He designed a circuit which could add binary numbers and later designed circuits which could make comparisons and thus is capable of performing a conditional statement. *This was the birth of digital logic and the digital computing age*.

He moved to the Mathematics Department at Bell Labs in the 1940s and commenced work that would lead to the foundation of modern *Information Theory*. The fundamental problem in this field is to reproduce at a destination point, either exactly or approximately, the message that has been sent from a source point. The problem is that information may be distorted by noise, leading to differences

---

[1] Victor Shestakov at Moscow State University also proposed a theory of electric switches based on Boolean algebra around the same time as Shannon. However, his results were published in Russia in 1941 whereas Shannon's were published in 1937.

G. O'Regan, *Giants of Computing: A Compendium of Select, Pivotal Pioneers*,
DOI 10.1007/978-1-4471-5340-5_50, © Springer-Verlag London 2013

Fig. 50.1 Claude Shannon

between the received message and the message that was originally sent. He provided a mathematical definition and framework for Information Theory in *A Mathematical Theory of Communication* [Sha:48].

He proposed the idea of converting data (e.g. pictures, sounds or text) to binary digits, i.e. binary bits of information. The information is then transmitted over the communication medium. Errors or noise may be introduced during the transmission, and the objective is to reduce and correct them. The received binary information is then converted back to the appropriate medium.

Shannon's theory was an immediate success with communications engineers. He also contributed to the field of cryptography in *Communication Theory of Secrecy Systems* [Sha:49].

He also made contributions to genetics and invented a chess-playing computer program in 1948. He built some early robot automata, game playing devices and problem-solving machines. He was able to juggle while riding a unicycle. He received many honours and awards and died in 2001.

50.1   Boolean Algebra and Switching Circuits

Vannevar Bush (discussed in an earlier chapter) was Shannon's supervisor at MIT, and Shannon's initial work was to improve Bush's mechanical computing device known as the differential analyzer.

This machine had a complicated control circuit that was composed of 100 switches that could be automatically opened and closed by an electromagnet. Shannon's insight was his realization that an electronic circuit is similar to Boolean algebra, and he showed how Boolean algebra could be employed to optimize the design of systems of electromechanical relays used in the analog computer. He also realized that circuits with relays could solve Boolean algebra problems.

Fig. 50.2 Switching circuit
representing Boolean logic

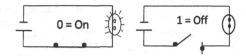

He showed in his Master's thesis "A symbolic analysis of relay and switching circuits" [Sha:37] that the binary digits (i.e. 0 and 1) can be represented by electrical switches.

The implications of true and false being denoted by the binary digits one and zero were enormous, since it allowed binary arithmetic and more complex mathematical operations to be performed by relay circuits. This provided electronics engineers with the mathematical tool they needed to design digital electronic circuits and provided the foundation of digital electronic design.

The design of circuits and telephone routing switches could be simplified with Boolean algebra. Shannon showed how to lay out circuitry according to Boolean principles, and his Master's thesis became the foundation for the practical design of digital circuits. These circuits are fundamental to the operation of modern computers and telecommunication systems, and his insight of using the properties of electrical switches to do Boolean logic is the basic concept that underlies all electronic digital computers (Fig. 50.2).

Digital circuits may be employed to implement Boolean algebra with the Boolean value 0 represented by a closed circuit, and the Boolean value 1 represented by an open circuit. A circuit may be represented by a set of equations with the terms in the equations representing the various switches and relays in the circuit. He developed a calculus for manipulating the equations, and this calculus is similar to Boole's algebra. The design of a circuit consists of algebraic equations, and the equations may be manipulated to yield the simplest circuit. The circuit may then be immediately drawn. Complex Boolean value functions can be constructed by combining these digital circuits. Next, we discuss Shannon's contribution to Information Theory.

50.2 Information Theory

Nyquist and others did early work on Information Theory at Bell Labs in the 1920s. Nyquist investigated the speed of transmission of information over a telegraph wire [Nyq:24] and proposed a logarithmic rule $(W = k \log m)^2$ that set an upper limit on the amount of information that may be transmitted. This rule is a special case of Shannon's later logarithmic law.

---

[2] W stands for the speed of transmission of information; m is the number of voltage levels to choose from at each step and k is a constant.

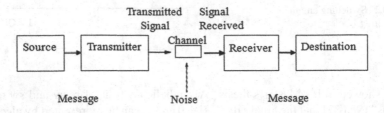

Fig. 50.3 Information theory

There were several communication systems in use prior to Shannon's 1948
paper. These included the telegraph machine, the telephone, the AM radio and early
television from the 1930s. These were all designed for different purposes and used
various media. Each of these was a separate field with its own unique problems,
tools and methodologies.

Shannon's classic 1948 paper [Sha:48] provided a unified theory for communication and a mathematical foundation for the field. The key problem in the field is
the reliable transmission of a message from a source point over a communications
channel to a destination point.[3] There may be noise in the channel that distorts the
message, and the engineer wishes to ensure that the message received is that which
has been sent.

The message may be in any communications medium, e.g. television, radio,
telephone and computers. Information Theory provides answers as to how rapidly
or reliably a message may be sent from the source point to the destination point.
Shannon identified five key parts of an information system (Fig. 50.3):

– Information source
– Transmitter
– Channel
– Receiver
– Destination

He derived formulae for the information rate of a source and for the capacity
of a channel including noiseless and noisy cases. These were measured in bits per
second, and he showed that for any information rate R less than the channel capacity
C,[4] it is possible (by suitable encoding) to send information at rate R, with an error
rate less than any pre-assigned positive ε, over that channel.

Shannon's theory of information is based on probability theory and statistics.
One important concept is that of *entropy*[5] which measures the level of uncertainty

---

[3]The system designer may also place a device called an encoder between the source and the channel
and a device called a decoder between the output of the channel and the destination.

[4]The channel capacity C is the limiting information rate (i.e. the least upper bound) that can be
achieved with an arbitrarily small error probability. It is measured in bits per second.

[5]The concept of entropy is borrowed from the field of thermodynamics.

in predicting the value of a random variable. For example, the toss of a fair coin has maximum entropy, as there is no way to predict what will come next. A single toss of a fair coin has entropy of one bit.

The concept of entropy is used by Shannon as a measure of the average information content missing when the value of the random variable is not known. English text has fairly low entropy as it is reasonably predictable because the distribution of letters is far from uniform.

Shannon proposed two important theorems that establish the fundamental limits on communication. The first theorem deals with communication over a noiseless channel and the second theorem deals with communication in a noisy environment.

The first theorem (*Shannon's source coding theorem*) essentially states that *the transmission speed of information is based on its entropy or randomness*. It is possible to code the information (based on the statistical characteristics of the information source) and to transmit it at the maximum rate that the channel allows. Shannon's proof showed that an encoding scheme exists, but did not show how to construct one. This result was revolutionary as communication engineers at the time thought that the maximum transmission speed across a channel was related to other factors and not on the concept of information.

Shannon's *noisy-channel coding theorem* states that reliable communication is possible over noisy channels provided that the rate of communication is below a certain threshold called the *channel capacity*. This result was revolutionary as it showed that a *transmission speed arbitrarily close to the channel capacity could be achieved with an arbitrarily low error*. The assumption at the time was that the error rate could only be reduced by reducing the noise level in the channel. Shannon showed that the desired transmission speed could be achieved by using appropriate encoding and decoding systems.

Shannon's theory also showed how to design more efficient communication and storage systems.

50.3 Cryptography

Shannon is considered the father of modern cryptography with his influential 1949 paper "Communication theory of secrecy systems" [Sha:49]. He established a theoretical basis for cryptography and defined the basic mathematical structures that underlie secrecy systems.

A secrecy system is defined to be a transformation from the space of all messages to the space of all cryptograms. Each possible transformation corresponds to encryption with a particular key, and the transformations are reversible. The inverse transformation allows the original message to be obtained provided that the key is known. A basic secrecy system is described in Fig. 50.4:

The first step is to select the key and to send it securely to the intended recipient. The choice of key determines the particular transformation to be used, and the message is then converted into a cryptogram (i.e. the encrypted text).

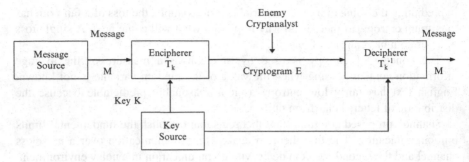

Fig. 50.4 Cryptography

The cryptogram is then transmitted over a channel (not necessarily secure) to the receiver, and the recipient uses the key to apply the inverse transformation to decipher the cryptogram into the original message.

The enciphering of a message is a functional operation. Suppose M is a message, K the key and E is the encrypted message, then

$$E = f(M, K)$$

This is often written as a function of one variable $E = T_i M$ (where the index i corresponds to the particular key being used). It is assumed that there are a finite number of keys $K_1, \ldots K_m$ and a corresponding set of transformations $T_1, T_2, \ldots, T_m$. Each key has a probability p_i of being chosen as the key. The encryption of a message M with key K_i is therefore given by

$$E = T_i M$$

It is then possible to retrieve the original message from the received encrypted message by

$$M = T_i^{-1} E$$

The channel may be intercepted by an enemy who will examine the cryptogram and attempt to guess the key to decipher the message. Shannon also showed that Vernam's cipher (also known as the *one time pad*) is an unbreakable cipher, and this cipher was invented by Gilbert Vernam at Bell Labs.

Chapter 51
William Shockley

William Shockley is famous for his invention of the transistor, and his contributions to the semiconductor field. The transistor was invented at Bell Labs by Shockley, Bardeen and Brattain, and they were awarded the Nobel Prize in physics in 1956 for their invention. Shockley later became the director of Shockley Semiconductor Laboratory, and he was involved in research and development of new transistors and other semiconductor devices (Fig. 51.1).

He was born in England in 1910 to American parents and grew up in Palo Alto in California. He obtained a Bachelor of Science degree from California Institute of Technology in 1932 and earned his Ph.D. from Massachusetts Institute of Technology in 1936. He joined Bell Labs shortly afterwards and remained there (apart from absences during the Second World War) until 1955. He published several papers on solid-state physics at Bell Labs.

He became involved in radar research at the start of the Second World War, and he took leave of absence from Bell Labs to become the research director at Columbia University's Anti-Submarine Warfare Operations Group in 1942. This involved determining methods to reduce the threat to convoys in the Atlantic from enemy submarines. He served as an expert consultant in the office of the Secretary of War, and he was involved in calculating the number of casualties that would be incurred in a land invasion of Japan.

He returned to Bell Labs at the end of the war and set up the solid physics research group. This group included Bardeen and Brattain, and this group developed the first transistor. They were awarded the Nobel Prize in physics for this invention.

He resigned from Bell Labs in 1955 and became director of Shockley Laboratory for Semiconductors at Mountain View in California. This company played an important role in the development of transistors and semiconductors, and several of its staff later formed semiconductor companies in the Silicon Valley area.

He became interested in the controversial field of eugenics in his later life, and he saw this work as important to the genetic future of the human race. He was criticized for these views, with one reporter comparing his views to Nazism, and he became

G. O'Regan, *Giants of Computing: A Compendium of Select, Pivotal Pioneers*,
DOI 10.1007/978-1-4471-5340-5_51, © Springer-Verlag London 2013

Fig. 51.1 William Shockley
(Courtesy Chuck Painter,
Stanford News Service)

isolated from friends and family. Others argue that his views were misrepresented in the media by journalists, and that other scholars who shared his views were reluctant to support him due to fear of being criticized themselves.

Shockley was granted over 90 patents during his lifetime and received several awards including the Medal of Merit in 1946. He received this for his work with the War Department during the Second World War. He received the Nobel Prize in physics with Bardeen and Brattain in 1956 for the invention of the transistor. He was a recipient of the Holley Medal of the American Society of Mechanical Engineers in 1963. He was the author of numerous articles and several books, and he died in 1989.

51.1   Invention of Transistor

Shockley's goal on his return to Bell Labs after the war was to find a solid-state alternative to the existing glass-based vacuum tubes. His research group included John Bardeen and Walter Brattain, and they would later receive the Nobel Prize in physics with him for their joint invention of the transistor.

Their early research was unsuccessful, but by late 1947 Bardeen and Brattain succeeded in creating a point-contact transistor independently of Shockley, who was working on a junction-based transistor. Bell Labs submitted several patent applications for the point-contact transistor, and Shockley was angered as his name did not appear on any of these applications. He believed that the point-contact transistor would not be commercially viable, and his junction point transistor was announced in mid-1951 with a patent granted later that year. The junction point transistor soon eclipsed the point-contact transistor and became dominant in the market place (Fig. 51.2).

Shockley published *Electrons and Holes in Semiconductors with Applications to Transistor Electronics* in 1950 [Sho:50]. This was the first textbook for scientists and engineers, who were interesting in learning about the new field of transistors and semiconductors.

Fig. 51.2 Replica of
transistor (Courtesy of Lucent
Bell Labs)

Shockley had a domineering personality, and he was difficult to work with at
times. His relationship with Bardeen, Brattain and the others deteriorated. He took
a position as visiting lecturer at California Institute of Technology in 1953.

51.2 Shockley Semiconductor Laboratory

He formed Shockley Semiconductor Laboratory (part of Beckman Instruments) in
1955. However, his management style soon alienated several of his employees. This
led to the resignation of eight key researchers in 1957 following his decision not to
continue research into silicon-based semiconductors.

This gang of eight went on to form Fairchild Semiconductors and other companies in the Silicon Valley area in the succeeding years. They included Gordon
Moore and Robert Noyce, who founded Intel in 1968. National Semiconductors
and Advanced Micro Devices were formed by other employees from Fairchild.
Shockley Semiconductors and these new companies formed the nucleus of what
became Silicon Valley.

The second generation of computers used transistors instead of vacuum tubes.
The University of Manchester's experimental Transistor Computer was one of the
earliest transistor computers. The prototype machine appeared in 1953 and the
full-size version was commissioned in 1955. The prototype machine had 92 point-
contact transistors.

Chapter 52
Richard Stallman

Richard Matthew Stallman is an American computer scientist who is famous as the *prophet of the free software movement*. He is president of the *Free Software Foundation* and has played a key role promoting the rights and freedom of software end users to use, copy and modify software. The existing intellectual property rights for software are quite stringent, and Stallman has sought ways to maximize freedom for software end users (Fig. 52.1).

He was born in New York in 1953 and became interested in computers while still at high school, after spending a summer working at IBM's Scientific Centre in New York. He obtained a degree in physics from Harvard University in 1974, and he joined the Artificial Intelligence Laboratory at MIT as a programmer. He became involved in the hacker community while at Harvard, and he later became a critic of restricted computer access at the lab.

He became convinced that software users should have the freedom to share software with others and to be able to study and make changes to the software that they use. He left his position at MIT in 1984 to launch a free software movement.

He launched the GNU project in 1984, which is a free software movement and involves the participation of software programmers from around the world. He formed the Free Software Foundation (FSF) to promote the free software movement. This organization has developed a legal framework for the free software movement, which provides a legal means to protect the modification and distribution rights of free software.

He has received several awards for his contribution to the computing field. These include the ACM Grace Murray Hopper Award, which he received in 1990 for pioneering developments of Emacs. He has received several honorary doctorates from universities around the world. He is the author of several books and technical papers, and selected essays of Richard Stallman are published in [Sta:02].

G. O'Regan, *Giants of Computing: A Compendium of Select, Pivotal Pioneers*, 243
DOI 10.1007/978-1-4471-5340-5_52, © Springer-Verlag London 2013

Fig. 52.1 Richard Stallman
(Courtesy of Victor Powell)

52.1 Free Software Foundation

Stallman started the non-profit organization, the Free Software Foundation (FSF), in 1985, and there are thousands of volunteer programmers involved. They develop free software as part of the free software movement. He is the non-salaried president of FSF, and the meaning of the term *free software* is defined in the GNU manifesto. He lists four key freedoms essential to software development [Sta:02], and a program is termed *free* if it satisfies these properties. These are:

1. Freedom to run the program for any purpose
2. Freedom to access, study and to improve the code, and to modify it to suit your needs
3. Freedom to make copies of the program and to redistribute them to others
4. Freedom to distribute copies of the modified program so that others can benefit from your improvements

The GNU project uses software that is free for users to copy, edit and distribute. It is free in the sense that users can change the software to fit individual needs. Stallman has written many essays on software freedom and is a key campaigner for the free software movement. The FSF has provided a legal framework for the free software movement to protect the modification and distribution rights of free software. Stallman introduced the important concept of *copyleft*, which is a form of *licensing of free software*. It makes a program or product free and requires that all modified or extended versions of the program are also free.

Copyright law grants exclusive rights to the copyright holder for a period of time to reproduce the work, to extend the work, to distribute copies of the work and to perform or display the work. The period of time may be 50 years after the author's death. Any unauthorized use of the author's work is termed a *copyright infringement*, and the copyright owner may take civil action to deal with the infringement.

A *patent is a form of intellectual property that grants exclusive rights to the inventor* to the commercial benefits of the invention for a limited period of time. The invention must be novel and more than just an obvious next step from existing inventions. The rights are granted to the inventor in a particular country, and typically for a period of 20 years. The rights granted are exclusive and prevent others from the unauthorized use of the invention. Patents may be legally enforced by civil lawsuits, and *patent licensing agreements* are legal contracts where the patent owner grants rights to others to use the invention.

Stallman has argued against intellectual property such as patent law and copyright law. He has argued against patenting software ideas, stating that a patent is an absolute monopoly on the use of an idea. He states that while 20 years may not seem like a long period of time, that in the software field it is essentially a generation, due to the pace at which technology changes in the world we live in. Further, *patents act as a barrier to competition and lead to monopolies.* They make it difficult for new companies to enter a market place, due to the restrictions and costs associated with the licensing of patents. In recent times, we have seen large companies acquire others for their intellectual property (e.g. the Google acquisition of Motorola Mobility was due to the latter's valuable collection of patents), and today there are major intellectual property wars in the corporate world.

Stallman argues that copyright law places Draconian restrictions on the general public and takes away freedoms that they would otherwise have. They protect the copyright owner businesses, and he suggests that the digital era requires us to consider alternative approaches.

52.2 GNU

GNU has a recursive definition and it stands for GNU's Not UNIX. The founding goal of the GNU project was to develop *a sufficient body of free software to get along without any software that is not free.* Stallman announced his plan to create a free operating system called GNU in 1983, and he published the GNU manifesto in 1985. The manifesto aimed to gain support from other developers for the project, and the initial goal was the development of a new operating system that would be compatible with UNIX.

He created several tools for GNU, including the Emacs text editor, the GCC compiler, the gdb debugger and the gmake build automator. A Finnish student, Linus Torvalds, used the GNU tools to create the Linux kernel in 1992, and the resulting free operating system platform is known as GNU Linux, or just Linux.

Chapter 53
Bjarne Stroustrup

Bjarne Stroustrup is a Danish computer scientist who is famous for designing and developing the C++ programming language. He is the author of several books on C++ and has many other publications. C++ is a widely used object-oriented language (Fig. 53.1).

He was born in Aarhus, Denmark, in 1950 and studied mathematics and computer science at Aarhus University. He received his Master's degree in mathematics and computer science in 1975. He earned his Ph.D. degree in computer science from the University of Cambridge in 1979. His Ph.D. was concerned with the design of distributed systems.

He moved to New Jersey in 1979 and joined the Computer Science Research Center at Bell Labs (later to become AT&T Bell Labs). He was the head of the Large-Scale Programming Research Department from its creation until 2002, when he moved to the University of Texas. He currently works as a distinguished professor of computer science at the University of Texas.

He has received several awards for his contributions including the Grace Murray Hopper Award in 1993. He was named an AT&T fellow in 1996, an ACM fellow in 1993 and an IEEE fellow in 2005.

53.1 C++ Programming Language

Stroustrup developed the C++ programming language in 1983 as an object-oriented extension of the C programming language. It was designed to use the power of object-oriented programming and to maintain the speed and portability of C. It provides a significant extension of C's capabilities but does not force the programmer to use the object-oriented features of the language.

A key difference between C++ and C is in the concept of a class. A *class* is an extension to the concept of a structure which used in C. The main difference is that while a *C* data *structure* can hold only *data,* a *C++ class* may hold both *data*

G. O'Regan, *Giants of Computing: A Compendium of Select, Pivotal Pioneers,*
DOI 10.1007/978-1-4471-5340-5_53, © Springer-Verlag London 2013

Fig. 53.1 Bjarne Stroustrup

and *functions*. An *object* is an instantiation of a class; i.e. the class is essentially the type, whereas an object is essentially a variable of that type. Classes are defined in C++ by using the keyword *class* as follows:

```
class class_name
{
  access_specifier_1:
    member1;
  access_specifier_2:
    member2;
  ...
}
```

The members may be either data or function declarations, and an access specifier is used to specify the access rights for each member (e.g. *private*, *public* or *protected*). Private members of a class are accessible only by other members of the same class, public members are accessible from anywhere where the object is visible and protected members are accessible by other members of same class and also from members of their derived classes. This is illustrated in the example of the definition of the class rectangle:

```
class CRectangle
{
  int x, y;
    public:
      void set_values (int,int);
      int area (void);
} rect;
```

53.2 Object-Oriented Languages

Object-oriented programming is a paradigm shift in programming. The traditional view of programming is *procedural*, where a program is considered to be a collection of functions (i.e. a list of instructions to be performed on the computer). Object-oriented programming considers a program to be a collection of *objects* that act on each other. Each object is capable of sending and receiving messages and processing data and may be viewed as an independent entity/actor with a distinct role or responsibility.

An object is a *black box* which sends and receives *messages*. A black box consists of *code* (computer instructions) and *data* (information which these instructions operate on). The traditional way of programming kept code and data separate. The functions and data structures in the C programming language are separate, whereas in the object-oriented world of C++, code and data are merged into a single indivisible thing called a *class* (and *objects* of the class).

The user of an object never needs to look inside the black box in order to use it, since all communication to it is done via *messages*. Messages define the *interface* to the object, and everything that an object can do is represented by its message interface. The approach of accessing an object only through its message interface, while keeping the internal details private is called *information hiding*[1] and dates back to work done by Parnas in the early 1970s.

The origins of object-oriented programming go back to the invention of Simula 67 at the Norwegian Computing Research Centre (NR)[2] in the late 1960s. Simula 67 introduced the notion of a class and instances of a class.[3] The Smalltalk language was developed at Xerox in the mid-1970s, and it introduced the term *object-oriented programming* for the use of objects and messages as the basis for computation. Many modern programming languages (e.g. Java and C++) support object-oriented programming. The main features of object-oriented languages are shown in Table 53.1.

Object-oriented programming became the dominant paradigm in programming from the late 1980s. Its proponents argue that it is easier to learn and simpler to develop and maintain. Its growth in popularity was helped by the rise in popularity of graphical user interfaces (GUI), as the development of GUIs is especially suited to object-oriented programming. The C++ language has become very popular, and object-oriented features have been added to many existing languages including COBOL and FORTRAN.

---

[1] Information hiding is a key contribution by Parnas to computer science. He has also done work on mathematical approaches to software quality using tabular expressions [ORg:06].

[2] The inventors of Simula 67 were Ole-Johan Dahl and Kristen Nygaard.

[3] Dahl and Nygaard were working on ship simulations and were attempting to address the huge number of combinations of different attributes from different types of ships. Their insight was to group the different types of ships into different classes of objects, with each class of objects being responsible for defining its own data and behaviour.

Table 53.1 Object-oriented paradigm

Feature	Description
Class	A class defines the abstract characteristics of a thing, including its attributes (or properties) and its behaviours (or methods). The members of a class are termed objects
Object	An object is a particular instance of a class with its own set of attributes. The set of values of the attributes of a particular object is called its state
Method	The methods associated with a class represent the behaviours of the objects in the class
Message passing	Message passing is the process by which an object sends data to another object or asks the other object to invoke a method
Inheritance	The programmer creates new classes (*subclasses*) from existing classes. Subclasses (or children classes) are more specialized versions of the class, and the derived classes inherit the methods and data structures of the parent class
Encapsulation (information hiding)	The internals of an object are kept private to the object and may not be accessed from outside the object. The details of how a particular class works are hidden, and a clearly specified interface around its services is provided
Abstraction	Abstraction simplifies complexity by modelling classes and removing all unnecessary detail
Polymorphism	Polymorphism is a behaviour that varies depending on the class in which the behaviour is invoked. Two or more classes may react differently to the same message

Chapter 54
Alan Turing

Alan Mathison Turing was a British mathematician and computer scientist who made fundamental contributions to mathematics and computer science. These include contributions to computability with his theoretical Turing machine, cryptography and breaking the Enigma codes at Bletchley Park during the Second World War, the design of the ACE machine at the National Physical Laboratory (NPL), the development of software for the Manchester Mark I and contributions to the emerging field of artificial intelligence (Fig. 54.1).

He was born in London 1912 and his father worked in the Indian Civil Service. He attended the Sherborne School[1] which was a famous public school in Dorset, England. Turing's interests at school were in science, mathematics and chess rather than the classics.

He excelled at long-distance running at the Sherborne, and in later life he completed the marathon in 2 h and 46 min. This time would have made him a candidate for the 1948 Olympic Games that were held in Wembley Stadium in London but he was injured before the games.

He attended King's College, Cambridge, from 1931 to 1934 and graduated with a distinguished degree in mathematics. He was elected a fellow of the college in 1935, and the computer room at King's College is named after him. He published a key paper in 1936 on a theoretical machine known as the "Turing machine", and he proved that anything that is computable is computable by this theoretical machine. This is known as the *Church-Turing thesis*.

He contributed to the code-breaking work carried out at Bletchley Park during the Second World War. The team at Bletchley succeeded in breaking the German Enigma codes. After the war he designed the ACE machine at the National Physical Laboratory and later worked on the Manchester Mark I computer. He did important work in artificial intelligence and devised the famous *Turing Test*, which is a test of machine intelligence.

---

[1] The Sherborne is located in Dorset, England. Its origins go back to the eight century when the school was linked with the Benedictine Abbey in the town.

G. O'Regan, *Giants of Computing: A Compendium of Select, Pivotal Pioneers*,
DOI 10.1007/978-1-4471-5340-5_54, © Springer-Verlag London 2013

Fig. 54.1 Alan Turing

Turing was a homosexual which was illegal in England at the time. He had a brief relationship with Arnold Murray in the early 1950s, and Murray and an accomplice burgled Turing's house shortly afterwards. Turing reported the matter to the police and allegations of homosexuality were made against him at the trial. He was charged, convicted and given a choice between imprisonment and probation. The terms of the probation were severe, and he received oestrogen hormone injections for a year. The side effects of the treatment led to depression. His security clearance was removed leading to the end of his work on cryptography and early computers. He committed suicide in June 1954.

Turing was awarded the Order of the British Empire (OBE) in 1945 in recognition of his wartime services to Britain. The Association for Computing Machinery awards the annual *Turing Award* to individuals who have made outstanding contributions to the computing field. He received a posthumous apology from the British government in 2009.

54.1   Turing Machines and Computability

Turing introduced the theoretical Turing machine in 1936. It consists of a head and a potentially infinite tape that is divided into frames. Each frame may be either blank or printed with a symbol from a finite alphabet of symbols. The input tape may initially be blank or have a finite number of frames containing symbols. At any step, the head can read the contents of a frame; the head may erase a symbol on the tape, leave it unchanged or replace it with another symbol. It may then move one position to the right, one position to the left or not at all. If the frame is blank, the head can either leave the frame blank or print one of the symbols.

Turing believed that every calculation could be done by a human with finite equipment and with an unlimited supply of paper to write on. The unlimited supply of paper is formalized in the Turing machine by a paper tape marked off in squares. The finite number of configurations of the Turing machine was intended to represent the finite states of mind of a human calculator (Fig. 54.2).

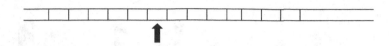

Fig. 54.2 Potentially infinite tape

The Turing machine is equivalent to an actual physical computer in the sense that it can compute exactly the same set of functions. It is much easier to analyze and prove things about than a real computer, but it is not suitable for programming and therefore does not provide a good basis for studying programming and programming languages.

The actions on a Turing machine are defined by the transition function, and it may be programmed to solve any problem for which there is an algorithm. However, if the problem is unsolvable, then the machine will either stop or compute forever. The solvability of a problem may not be determined beforehand. There is some answer (i.e. either the machine halts or it computes forever).

Turing showed that there was no solution to the *Entscheidungsproblem* (i.e. the *decision problem*) posed by the German mathematician David Hilbert. This problem asserted that the truth or falsity of a mathematical problem can always be determined by a mechanical procedure. Hilbert was asserting that first-order logic is decidable; i.e. there is a decision procedure to determine if an arbitrary sentence in the formula is valid. The Turing machine played a key role in refuting Hilbert's claim of the decidability of first-order logic.

Turing had already proved that the halting problem for Turing machines is not computable; i.e. it is not possible algorithmically to decide whether or not any given Turing machine will halt or not. The question as to whether a given Turing machine halts or not can be formulated as a first-order statement. If a general decision procedure exists for first-order logic, then the statement of whether a given Turing machine halts or not is within the scope of the decision algorithm. Therefore, since there is no general algorithm that can decide whether any given Turing machine halts, there is no general decision procedure for first-order logic. The only way to determine whether a statement is true or false is to try to solve it; however, if one tries but does not succeed, this does not mean that an answer does not exist.

54.2 Bletchley Park

The team at Bletchley Park played an important role in breaking the Enigma codes which contributed to the defeat of Nazi Germany during the Second World War. These codes were used by the Germans in the transmission of naval messages, to their submarines, and were designed to ensure that any unauthorized interception of the messages would be unintelligible to the third party (Fig. 54.3).

Fig. 54.3 Bletchley Park

Fig. 54.4 The Enigma
machine (Public Domain)

The plaintext (i.e. the original message) was converted by the Enigma machine
into the encrypted text, and these messages were then transmitted by the Germans
to their submarines in the Atlantic or to their bases throughout Europe (Fig. 54.4).

The Enigma codes were cracked by a team of cryptanalysts at Bletchley Park,
using a machine called the Bombe. The Poles had done some work on code breaking

Fig. 54.5 Rebuilt Bombe (Photo Public Domain)

prior to the war, and they passed their knowledge to the British following the German invasion of Poland. Turing and Gordon Welchman built on the Polish research to develop the *Bombe machine*. They observed that a message often contained common words or phrases, such as a general's name or weather reports. This enabled them to guess short parts of the original message. These guesses were called *cribs*.

The Enigma machine does not allow a letter to be enciphered as itself, and this reduced the potential number of settings that the Enigma machine could be in on that particular day. The code-breaking team then wired the Bombe to check the reduced set of settings. The Bombe found potential Enigma settings not by proving a particular setting, but by disproving every incorrect one in turn (Fig. 54.5).

The first Bombe was installed in early 1940, and there were over 200 of them in operation by the end of the war. They were built by the British Tabulating Machine Company, and a replica of the British Bombe machine was rebuilt at Bletchley Park by a team of volunteers in 2008.

54.3 National Physical Laboratory

Turing worked on the design of the Automatic Computing Engine (ACE) at the National Physical Laboratory (NPL) after the war. This was intended to be a smaller and more advanced computer than ENIAC. His design included detailed

Fig. 54.6   NPL Pilot ACE (Courtesy of NPL © Crown Copyright 1950)

logical circuit diagrams and he proposed a high-speed memory. The machine would implement subroutine calls, which set it apart from the EDVAC, and it used an early form of programming language termed *Abbreviated Computer Instructions*. The estimated cost was 11,000. Turing's colleagues at NPL thought that his design was too ambitious.

There were delays with funding and Turing became disillusioned with NPL. His original machine was never built and NPL built a smaller version called the ACE Pilot Model. It ran its first program in 1950 and it was then the fastest computer in the world with a clock speed of 1 MHz. Turing moved to the University on Manchester to work on the Manchester Mark I computer (Fig. 54.6).

54.4   Turing Test in AI

Turing contributed to the debate concerning artificial intelligence in his paper on *computing, machinery and intelligence* [Tur:50], which explored whether it would be possible for a machine to be conscious and to think. He devised a famous experiment (known as the *Turing Test*) that would determine whether a computer was intelligent or not.

The test is an adaptation of a party game, which involves three participants. One of them, the judge, is placed in a separate room from the other two: one is male and the other is female. Questions and responses are typed and passed under the door.

The objective of the game is for the judge to determine which participant is male and which is female. The male is allowed to deceive the judge whereas the female is supposed to assist.

Turing adapted this game to allow a computer to play the role of the male. The computer is said to pass the Turing Test if the judge is unable to determine which of the participants is human and which is machine. His paper was controversial, as defenders of traditional values attacked the idea of machine intelligence. Turing strongly believed that machines would eventually be developed that would stand a good chance of passing the test.

The viewpoint that a machine will one day pass the Turing Test and be considered intelligent is known as *strong AI*. It states that a computer with the right program would have the mental properties of humans. There are a number of objections to strong AI, and one well-known rebuttal is *Searle's Chinese Room* argument [Sea:80].

Chapter 55
Thomas Watson Sr. and Jr.

Thomas Watson Sr. and Jr. are famous past presidents of International Business Machines (IBM). Thomas Watson Sr. transformed IBM into an international company that sold punched card tabulating machines, and Thomas Watson Jr. transformed IBM to an international computer company building and selling computers around the world. IBM became the dominant player in the computer industry.

55.1 Thomas Watson Sr.

Thomas Watson Sr. was born in New York in 1874. He worked as a salesman for NCR and later ran the NCR Agency in Rochester, New York. He became embroiled in an antitrust affair taken by the US government against NCR, with allegations of questionable anticompetitive business practices employed by NCR. He was eventually acquitted, and he joined the Computer Tabulating Recording Company (CTR) as general manager in 1914. He renamed the company to International Business Machines (IBM) in 1924 (Fig. 55.1).

He transformed the company into a highly successful international company based around punched card tabulating machines. He was responsible for the famous *THINK* signs that have been associated with IBM for many years. They were introduced in 1915.

Watson considered the motivation of the sales force to be an essential part of his job, and the sales people were required to attend an *IBM sing along*. The verses in the songs were in praise of IBM and its founder Thomas Watson Sr. These songs were published as a book *Songs of the IBM* in 1931 and included *Ever Onward*, *March on with IBM* and *Hail to the IBM*.

IBM developed the popular IBM punched card in the late 1920s which provided almost double the capacity of existing cards. The company introduced a mechanism by which staff could make improvement suggestions in the 1920s.

G. O'Regan, *Giants of Computing: A Compendium of Select, Pivotal Pioneers*,
DOI 10.1007/978-1-4471-5340-5_55, © Springer-Verlag London 2013

Fig. 55.1 Thomas
Watson Sr.

The great depression in the United States did not have a major impact on IBM, and IBM's policy was to take care of its employees. It was one of the first corporations to provide life insurance and paid vacations for its employees. Watson kept his workers busy during the depression by producing new machines even while demand was slack. He also won a major government contract to maintain employment records for over 26 million people.

He recognized the importance of research and development and created a division in the early 1930s to lead the engineering, research and development efforts for the entire IBM product line. It recognized the importance of education and development of employees to the success of its business. IBM launched an employee and customer magazine called *Think* in the 1930s, and this magazine included topics such as education and science.

IBM placed all of its plants at the disposal of the US government during the Second World War, and it expanded its product line to include military equipment. It commenced work on computers during the war years with the Harvard Mark I (also known as the *IBM Automatic Sequence Controlled Calculator* (ASCC)). The machine was designed by Howard Aiken and built by IBM. It was essentially an electromechanical calculator that could perform large computations automatically. It was completed in 1944 and presented to Harvard University.

The machine was designed to assist in the numerical computation of differential equations, and it was funded and built by IBM. It was over 50 ft long, 8 ft high and weighed 5 tonnes. It performed additions in less than a second, multiplications in 6 s and division in about 12 s. It used electromechanical relays to perform the calculations.

IBM employed 1,300 people and had $9 million of revenue when Watson took over. Its revenue was $897 million and it employed over 72,000 people when he died in 1956.

55.2   Thomas Watson Jr.

Thomas Watson Jr. was an American businessman and philanthropist, and he served as president of IBM from 1952 to 1971. He transformed IBM from a company that was focused on tabulators to a company that became the leader in the emerging computer technology field (Fig. 55.2).

He was born in 1914 and obtained a business degree from Brown University. He joined IBM as a salesman and served as a pilot in the US Air Force during the Second World War.

Thomas Watson Sr. retired in 1952 and Thomas Watson Jr. became chief executive officer the same year. He believed that the future of IBM was in computers rather than in tabulators. He recognized the future role that computers would play in business and realized that IBM needed to change to adapt to the new technology. He played a key role in the transformation of IBM to a company that would become the world leader in the computer industry.

He served as the American ambassador to the Soviet Union from 1979 to 1981. He received many awards including the Presidential Medal of Freedom from President Lyndon Johnson in 1964. He died of a stroke in 1993.

Fig. 55.2   Thomas Watson Jr.

5.8.2 Thomas Watson Jr.

Fig. 5.3 Thomas Watson

Chapter 56
Joseph Weizenbaum

Joseph Weizenbaum was a German-American computer scientist who is famous for his development of the ELIZA program in 1966 and for his views on the ethics of artificial intelligence. He became sceptical of artificial intelligence and a leading critic of the AI field following the response of users to the ELIZA program. Many users felt that they were communicating with an empathic psychologist rather than a machine. He was professor emeritus of computer science at MIT, and he also held academic positions at several other universities including Harvard, Stanford, the Technical University of Berlin and the University of Hamburg (Fig. 56.1).

He was born to Jewish parents in Berlin in 1923, and his family fled from Nazi Germany in 1936 and immigrated to Detroit in the United States. He began studying mathematics at Wayne University in the USA in 1941, but the Second World War interrupted his studies. He served in the military during the war as a meteorologist in the Army Air Corps and resumed his studies after the war. He was awarded an MS in mathematics in 1950.

He joined General Electric in 1955 and was part of the team that designed and implemented the first computerized banking system in the United States. He developed a list processing system called SLIP (Symmetric List Processor) in 1963. He became professor of computer science at MIT in 1963.

He used SLIP to write the ELIZA program at MIT in 1966, and this was one of the earliest AI programs and an important milestone in the AI field. This famous natural language understanding program was named after the character *Eliza.* in the musical *My Fair Lady*, which was based on Shaw's 1912 play *Pygmalion*. The ELIZA program simulated a conversation between a patient and a psychotherapist, with the program using the person's response to shape its reply. The interaction was between the computer program and a user sitting at an electric typewriter, with the user typing and the computer program responding. The program convinced several users that it had real understanding and that it was an empathic psychotherapist. This led to users unburdening themselves in long computer sessions. ELIZA was a sensation at the MIT campus, and it rapidly spread to other universities.

G. O'Regan, *Giants of Computing: A Compendium of Select, Pivotal Pioneers*,
DOI 10.1007/978-1-4471-5340-5_56, © Springer-Verlag London 2013

Fig. 56.1 Joseph
Weizenbaum

Weizenbaum was shocked to discover that so many users were taking his program
seriously and were sharing their most private thoughts with the machine. He became
an advocate of social responsibility in science and a leading critic of AI research.

He took 2 years leave of absence from MIT in 1972 to write a critique of AI,
and this critique is published in his influential book *Computer Power and Human
Reason* [Wei:76]. This popular and accessible book became a bestseller and was
translated into several languages. It deals with the role of computers in society. He
argues that while artificial intelligence may someday be possible, computers should
never be given responsibility for making important human decisions, as they lack
human qualities such as compassion and wisdom. The book raised questions on the
role of artificial intelligence and led to debate on the role of computer systems in
decision making and on the reliance on computer systems and technology in the
world today.

Weizenbaum received several awards in recognition of his contributions. He
was the recipient of CPSR's Norman Wiener Award for Professional and Social
Responsibility in 1988. He moved to Berlin in 1996 and died of cancer there in
2008.

56.1   Eliza

Weizenbaum created the famous ELIZA program at MIT in 1966, and the program
interacted with a user sitting at an electric typewriter, in English, in the manner of a
Rogerian psychotherapist.[1]

---

[1] Rogerian psychotherapy is a form of psychotherapy developed by Carl Rogers in the 1940s and
1950s. It is also known as person-centred therapy.

The name *Eliza*[2] was chosen because it could be taught to speak increasingly well, as in Shaw's Pygmalion. It was one of the earliest natural language processing programs, and Weizenbaum gave an example of a conversation between Eliza and a young woman in [Wei:66]. Many users believed that Eliza had real understanding, and they began to unburden themselves in long computer sessions.

The program operated by decomposing the user's input into its constituent parts of speech and then typing them back at the user in a manner to continue with the dialogue. The program provided very convincing human-like interaction, as may be seen in the sample dialogue below:

User/patient:	Men are all alike.
ELIZA:	*In what way?*
User/patient:	They are always bugging us about something or other.
ELIZA:	*Can you think of a specific example?*
User/patient:	Well, my boyfriend made me come here.
ELIZA:	*Your boyfriend made you come here?*
User/patient	He says I'm depressed much of the time.
ELIZA:	*I am sorry to hear that you are depressed.*
User/patient:	It's true, I am unhappy.
ELIZA:	*Do you think that coming here will help you not to be unhappy?*
User/patient:	I need some help, that much seems certain.
ELIZA:	*What would it mean to you if you got some help?*
User/patient:	Perhaps, I could learn to get along with my mother.
ELIZA:	*Tell me more about your family.*

Weizenbaum was shocked to discover that so many users were convinced that the program had real understanding, and users spent hours sharing their personal problems with the program. It led Weizenbaum to think about the ethics and implications of the artificial intelligence field, and the ability of a relatively trivial program could deceive a naïve user to reveal personal information. He became a leading critic of the AI field and an advocate for professional and social responsibility.

56.2 Ethics of Artificial Intelligence

Weizenbaum's views on the ethics of AI are discussed in his book Computer Power and Human Reason [Wei:76]. He displays ambivalence towards computer technology and argues that it is dangerous and immoral to assume that computers could eventually take over any role, given sufficient processing power and appropriate programming.

---

[2]Eliza Doolittle was a working class character in Shaw's play Pygmalion. She is taught to speak with an upper-class English accent by Professor Henry Higgins.

He argues that AI is a threat to human dignity, and that AI should not replace humans in positions that require respect and care. He states that machines lack empathy and that if they replace humans in positions such as police officers or judges, this would lead to alienation and a devaluation of the human condition. Weizenbaum's views have been criticized by John McCarthy and others.

Weizenbaum's ELIZA program demonstrated the threat that AI poses to privacy. It is conceivable that an AI program may be developed in the future that is capable of understanding speech and natural languages. Such a program could theoretically eavesdrop on every phone conversation and email and gather private information on what is said and who is saying it. Such a program could be used by a state to suppress dissent and to eliminate those who pose a threat to the state.

As more and more sophisticated machines and robots are created, it is, of course, essential that intelligent machines behave ethically and have a moral compass to distinguish right from wrong. It remains an open question as to how to teach a robot right from wrong, and the science fiction writer, Asimov, proposed three laws of robotics in his book, *I Robot*. These laws basically state that a robot must not injure a human being through either action or inaction. A robot is required to obey orders (provided that the order does not involve harm to another human being), and finally a robot must protect its own existence unless of course its continued existence will cause harm to humans.

Chapter 57
Frederick Williams

Sir Frederick Williams is famous for his work with Tom Kilburn on the use of cathode-ray tubes as an information storage device. This was known as the *Williams-Kilburn tube*, and it was the first form of random access memory. They then developed the *first stored program digital electronic computer* (the Manchester "Baby") in 1947 (Fig. 57.1).

He was born in Cheshire, England, in 1911 and did a degree in electrical engineering at the University of Manchester. He obtained a doctorate (DPhil) from Oxford in 1936.

He worked at the Telecommunications Research Establishment during the Second World War and was recognized as a world expert on radar. He moved to Manchester University in 1946 and became head of the Electrical Engineering Department.

Williams and Kilburn developed the Williams-Kilburn tube in 1946 to store binary data. This was the first random access digital storage device, and it was used successfully on several early computers. They developed the small-scale experimental machine (SSEM) in 1947, and this was the first stored program computer. It was also known as the Manchester "Baby" computer.

Queen Elizabeth knighted him in 1976 in recognition of his contributions to electronics and electrical engineering. He died in 1977.

57.1 Williams Tube

Williams and Tom Kilburn invented the Williams-Kilburn tube (also known as the Williams tube) in 1946. This was a cathode-ray tube used to store binary data, and each Williams tube could store 512–1024 bits of data. It was the first random access digital storage device, and it remained popular in the computer data-storage field for several years until it was outdated by core memory in 1955. It provided the first large amount of random access memory (RAM), and it was used as a key component in the first stored program computer.

G. O'Regan, *Giants of Computing: A Compendium of Select, Pivotal Pioneers*,
DOI 10.1007/978-1-4471-5340-5_57, © Springer-Verlag London 2013

Fig. 57.1 Sir Frederic
Williams

Williams had succeeded in storing one bit of information on a cathode-ray tube, and Kilburn began working with him in the mid-1940s to improve its digital storage ability. Kilburn devised an improved method of storing bits which increased the storage capacity. They were now ready to build a computer to test the reliability of the memory in the Williams tube, and this led to the Manchester "Baby" computer in 1948.

57.2 Manchester Baby

The Manchester Small-Scale Experimental Computer (better known by its nick-name "Baby") was developed at the University of Manchester. It was the *first stored program computer*, and it was designed and built at Manchester University in England by Frederic Williams, Tom Kilburn and others.

Kilburn was assisted by Geoff Tootill in the design and construction of the prototype machine, and the prototype machine demonstrated the reliability of the Williams tube. It was the first stored program computer; i.e. the instructions to be executed were loaded into memory rather than by rewiring the computer to run a new program. That is, all that was required was to enter the new program into the computer memory. Kilburn wrote and executed the first stored program, and it was a short 17-line program to computer the highest factor of 2^{18}. It was written and executed in 1948.

The prototype machine demonstrated the feasibility and potential of a stored program computer. Its memory consisted of 32 32-bit words, and it took 1.2 ms to execute one instruction, i.e. 0.00083 MIPS (million instructions per second). Today's computers are rated at speeds of up to 1,000 MIPS. The team in Manchester developed the machine further, and in 1949, the Manchester Mark I was available (Fig. 57.2).

Fig. 57.2 Replica of the Manchester Baby (Courtesy of Tommy Thomas)

57.3   Manchester Mark I

The Manchester Automatic Digital Computer (MADC), also known as the Manchester Mark I, was developed at the University of Manchester. It was one of the earliest stored program computers, and it was the successor to the Manchester "Baby" computer. It was designed and built by Williams, Kilburn and others.

Each word could hold one 40-bit number or two 20-bit instructions. The main memory consisted of two pages (i.e. two Williams tubes with each holding 32×40 bit words or 1280 bits). The secondary backup storage was a magnetic drum consisting of 32 pages (this was updated to 128 pages in the final specification). Each track consisted of two pages (2560 bits). One revolution of the drum took 30 ms, and this allowed the 2560 bits to be transferred to main memory.

It contained 4,050 vacuum tubes and had a power consumption of 25,000 W. The standard instruction cycle was 1.8 ms, but multiplication was much slower. The machine had 26 defined instructions, and the programs were entered into the machine in binary format, as assembly languages and assemblers were not yet available.

It had no operating system, and its only systems software were some basic routines for input and output. Its peripheral devices included a teleprinter and a 5-hole paper tape reader and punch.

A display terminal used with the Manchester Mark I computer mirrored what was happening within the Williams tube. A metal detector plate placed close to the

Fig. 57.3 The Manchester Mark I Computer (Courtesy of the University of Manchester)

surface of the tube detected changes in electrical charges. The metal plate obscured a clear view of the tube, but the technicians could monitor the tubes used with a video screen. Each dot on the screen represented a dot on the tube's surface, and the dots on the tube's surface worked as capacitors that were either charged and bright or uncharged and dark. The information translated into binary code (0 for dark, 1 for bright) became a way to program the computer (Fig. 57.3).

57.4 Ferranti Mark I

Ferranti Ltd. (a British company) and Manchester University collaborated to build one of the world's first commercially available general-purpose electronic computers. This was the Ferranti Mark I, and it was basically an improved version of the Manchester Mark I. The first machine off the production line was delivered to the University of Manchester in 1951, and shortly before the release of the UNIVAC 1 electronic computer in the United States.

The Ferranti Mark I's instruction set included a "hoot command" which allowed auditory sounds to be produced. It also allowed variations in pitch. Christopher Strachey (who later did important work on the semantics of programming languages) programmed the Ferranti Mark I to play tunes such as *God save the King*, and the Ferranti Mark I was one of the earliest computers to play music. The earliest was the CSIRAC in Australia. The parents of Tim Berners-Lee (the inventor of the World Wide Web) both worked on the Ferranti Mark I.

The main improvements of the Ferranti Mark I over the Manchester Mark I were improvements in size of primary and secondary storage, a faster multiplier and additional instructions.

It had 8 pages of random access memory (i.e. 8 Williams tubes each with a storage capacity of 64 20-bit words or 1280 bits). The secondary storage was provided by a 512-page magnetic drum which stored two pages per track, and its revolution time was 30 ms.

It used a 20-bit word stored as a single line of dots on the Williams tube display, with each tube storing a total of 64 lines of dots (or 64 words). Instructions were stored in a single word, while numbers were stored in two words.

The accumulator was 80 bits, and it could also be addressed as two 40-bit words. There were about 50 instructions and the standard instruction time was 1.2 ms. Multiplication could be completed in 2.16 ms. There were 4,050 vacuum tubes employed.

Chapter 58
Niklaus Wirth

Niklaus Wirth has made important contributions to software engineering and to the design of programming languages. He published the paper "Program development by stepwise refinement" and has designed and developed several programming languages including Pascal, Modula-2 and Oberon (Fig. 58.1).

He was born in Switzerland in 1934, and he obtained a bachelor's degree in electronic engineering from the Swiss Federal Institute of Technology (ETH) in 1959. He obtained his Ph.D. degree in electrical engineering and computer science from Berkeley University in California in 1963.

He held an assistant lecturer position at Stanford University from 1963 until 1967, and he then returned to Switzerland to a position at the University of Zurich. He was appointed professor of Informatics at ETH, Zurich in 1968 and remained there until his retirement in 1999.

He spent a sabbatical year at Xerox PARC in California in the mid-1970s and also in the mid-1980s. Wirth has made important contributions to software engineering and to the design and development of programming languages.

He designed and developed the Pascal programming language at Zurich from 1968 to 1972. This language was designed for use both as a teaching language and for practical software development.

Structured programming [Dij:68] is concerned with rigorous techniques to design and develop programs, and the debate on structured programming was quite intense in the late 1960s. The computing community was split into two professional camps, with scientists and engineers using FORTRAN for their large-scale scientific programming projects and the business using COBOL for their smaller applications. Pascal was an attempt to unite these two worlds, and to incorporate the concepts of structured programming and top-down design. The first Pascal compiler was available in 1970 on a mainframe computer, and compilers for other machines followed.

Modula-2 was designed in the mid-1970s as a successor to Pascal. Its objectives were to address weaknesses in Pascal and to deal with new challenges in the

G. O'Regan, *Giants of Computing: A Compendium of Select, Pivotal Pioneers*, DOI 10.1007/978-1-4471-5340-5_58, © Springer-Verlag London 2013

Fig. 58.1 Niklaus Wirth

programming field such as multiprogramming and information hiding. Wirth also desired that the language should be able to deal with the challenge of describing entire systems.

Modula-2 includes the module construct, and a module has an explicit interface specification to other modules. A module consists of two distinct parts: a definition part and an implementation part. The definition part consists of type and procedure signatures, and these are visible to other modules. The implementation part is hidden, and it consists of all local, hidden objects and the bodies of procedures. Modula-2 supports a modular decomposition of the system and the definition of the interfaces of the system. Modular decomposition and information hiding are due to Parnas, who was discussed in an earlier chapter. Modula-2 also included certain low-level features mainly to access particular machine resources, and this was done to make the language sufficiently powerful to describe entire systems.

Wirth designed the Oberon language with the goal to simplify a programming language to its essentials, as well as including the features that were essential to object-oriented programming. The goal of the language was to shield the programmer from implementation details, and to allow the programmer to think exclusively in terms of the higher-level abstraction.

Wirth has received various awards for his contributions to the computing field. He received the ACM Turing Award in 1984 for developing a sequence of innovative computer languages including Euler, Algol-W, Pascal and Modula-2. He received the ACM SIGSOFT Outstanding Research Award in Software Engineering in 1999.

58.1   Pascal Programming Language

The Pascal programming language is named after *Blaise Pascal* (a seventeenth-century French mathematician and inventor). It was developed in the early 1970s, and it was influenced by the ALGOL programming language. The main goal of the language was to teach students structured programming.

It includes constructs such as the conditional *if statement*; the iterative *while*, *repeat* and *for* statements; the *assignment statement*; and the *case statement* (which is a generalized if statement). The statement in the body of the repeat statement is executed at least once, whereas the statement within the body of a while statement may never be executed.

The language has several reserved words (known as keywords) that have a special meaning, and these may not be used as program identifiers. The Pascal program that displays "Hello World" is given by

```
program HELLOWORLD (OUTPUT);

begin
  WRITELN ('Hello World!')
end.
```

Pascal includes several simple data types such as Boolean, Integer, Character and Reals. Its more advanced data types include arrays, enumeration types, ordinal types and pointer data types. It allows complex data types to be constructed from existing data types. Types are introduced by the reserved word *type*.

```
type
  c = record
        a : integer;
        b : char
      end;
```

Pascal includes a "pointer" data type, and this data type allows linked lists to be created by including a pointer type field in the record. The variable LINKLIST is a pointer to the data type B in the example below where B is a record.

```
type
  BPTR = ^B;
  B = record
        A : integer;
        C : BPTR
      end;

var
  LINKLIST : BPTR;
```

Pascal is a block structured language with programs structured into procedures and function blocks. These can be nested to any depth, and recursion is allowed. Each block has its own constants, types, variables and other procedures and functions which are defined within the scope of the block.

58.2   Program Development by Stepwise Refinement

Program development by stepwise refinement is concerned with the design and construction of a program in a sequence of refinement steps. In each step, a task is broken down into a number of subtasks. Each refinement in the description of the task may be accompanied by a refinement in the description of the data. The latter serves as the means of communication between the subtasks. Wirth's 1971 paper [Wir:71] shows how a program may be gradually developed by a sequence of refinement steps.

Each refinement step involves decomposing an instruction (or several instructions) of the given program into more detailed instructions. The successive decomposition or refinement is complete when all the instructions are expressed in terms of a computer programming language. The refinement to the program instructions will lead to refinements of the data, and so program refinement and data refinement often occur in parallel.

Each refinement step involves a design decision, and the programmer chooses a particular solution from a number of the alternatives available.

The Vienna Development Method (VDM) is a formal development method which uses stepwise refinement. It employs rules to verify each refinement step, and this allows the executable specification, i.e. the detailed code, to be obtained from the initial specification via refinement steps. Thus, we have a sequence $S = S_0, S_1, \ldots, S_n = E$ of specifications, where S is the initial specification and E is the final (executable) specification.

$$S = S_0 \sqsubseteq S_1 \sqsubseteq S_2 \sqsubseteq \ldots \sqsubseteq S_n = E$$

Retrieval functions enable a return from a more concrete specification to the more abstract specification.

Chapter 59
Ed Yourdon

Edward Nash Yourdon is an American pioneer in software engineering methodologies and one of the leading developers of the structured analysis and design methods in the 1970s. He has also made important contributions to object-oriented analysis and design methods. He is a well-known international computer consultant specializing in project management, software engineering methodologies and web technologies. He is the author of over 20 books in the computing field (Fig. 59.1).

He was born in 1944 and obtained a bachelor's degree in applied mathematics from Massachusetts Institute of Technology in 1965. He commenced his computing career in 1964 as a senior programmer with Digital Equipment Corporation. He developed the FORTRAN Maths library for the PDP-5 computer and wrote an assembler for the PDP-8 computer. He then worked as a project manager at General Electric and became an independent consultant specializing in software development and project management methodologies from the early 1970s.

He became one of the leading developers of the structured analysis and design methods. These are methods for analyzing and converting business requirements into specification and programs.

Structured analysis involves drawing diagrams to represent the information flow in the system, as well as showing the key processes at the level of granularity required. Data dictionaries are used to describe the data, and CASE tools are employed to support the drawing of diagrams and recording the objects drawn in a data dictionary. His contribution to object-oriented analysis and design in the late 1980s and early 1990s includes codeveloping the Yourdon/Whitehead method of object-oriented analysis and design and the Coad/Yourdon object-oriented methodology.

He founded an international consulting business, Yourdon Inc., in 1974 with the goal of providing education, publishing and consulting services to clients in software engineering technologies. His consulting business was very successful, and it trained over 250,000 people around the world in structured programming, structured analysis, structured design, logical data modelling and project management. It was sold in the mid-1980s and is now part of CGI Informatique.

G. O'Regan, *Giants of Computing: A Compendium of Select, Pivotal Pioneers*,
DOI 10.1007/978-1-4471-5340-5_59, © Springer-Verlag London 2013

Fig. 59.1 Edward Yourdon
(Courtesy of Edward
Yourdon)

He founded and published the influential *Cutter IT Journal* and was the editor of the journal. He is a fellow in business technology trends council for the Cutter Consortium. He is the author of over 500 technical articles and over 20 books and is a keynote speaker at major computer conferences around the world.

He has received several awards for his contributions to the computing field and was inducted into the Computer Hall of Fame in 1997.

59.1   Yourdon Structured Method

Structured analysis and design are methods of analyzing and converting business requirements into specifications and ultimately into computer programs. Several structured methods arose in the 1970s including structured programming developed by Dijkstra, stepwise design developed by Wirth, structured design developed by Yourdon and others and Jackson structured programming (JSP) developed by Jackson and so on.

These analysis, design and programming techniques were designed to address the problems with software development in the late 1960s. At that time there was little guidance on sound techniques to design and develop software, and there were no agreed techniques for specifying and documenting requirements and design. Structured analysis views the system from the point of view of data flowing through the system and uses data flow diagrams.

The Yourdon Structured method was developed in the 1980s and revised in the 1990s. The method supports two distinct phases, namely, analysis and design. The method includes three basic steps: feasibility study, essential modelling and implementation modelling.

The method requires the systems analyst to first draw a context diagram, and this is the top-level data flow diagram (DFD). This indicates the source, sink and boundaries of the system. Data flow diagrams were described in the discussion on Tom DeMarco in an earlier chapter.

The various users are then interviewed, and the events to which the proposed system must respond are identified. The documentation of the system is then prepared using techniques such as entity relationship diagrams (ERDs), DFDs and structured English. Various tools are employed to produce graphical models of the system.

There are several graphical techniques employed to given different views of the system being developed, and three orthogonal viewpoints of the system are presented. This leads to a three-dimensional description of the system.

Information

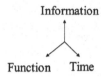

Function   Time

The information dimension represents the information used by the system and is described by entity relationship diagrams (ERD). The time dimension describes the events to which the system must respond. The function dimension describes what the system does and is described by data flow diagrams.

Chapter 60
Konrad Zuse

Konrad Zuse is considered *the father of the computer* in Germany, as he built the world's first programmable machine (the Z3) in 1941 (Fig. 60.1).

He was born in Berlin in 1910 and studied civil engineering at the Technical University of Berlin. He was talented at mechanical construction and won several prizes as a student for his constructions.

He commenced working as a stress analyzer for Henschel after his graduation in 1935. Henschel was an airline manufacturer, and Zuse stayed in his position for less than a year. He resigned with the intention of forming his own company to build automatic calculating machines.

His parents provided financial support, and he commenced work on what would become the Z1 machine in 1936. Zuse employed the binary system for the calculator and metallic shafts that could shift from position 0 to 1 and vice versa. The Z1 was operational by 1938.

He served in the German Army on the Eastern Front for 6 months in 1939 at the start of the Second World War. Henschel helped Zuse to obtain a deferment from the army and made the case that he was needed as an engineer and not as a soldier. Zuse recommenced working at Henschel in 1940 and remained affiliated to Henschel for the duration of the war. He built the Z2 and Z3 machines there. The Z3 was operational in 1941 and was the world's first programmable computer. He started his own company in 1941, and this was the first company founded with the sole purpose of developing computers. The Z4 was almost complete as the Red Army advanced on Berlin in 1945, and Zuse left Berlin for Bavaria with the Z4 prior to the Russian advance.

He designed the world's first high-level programming language between 1943 and 1945, and this language was called Plankalkül. He later restarted his company (Zuse KG), and he completed the Z4 in 1950. This was the first commercial computer as it was completed ahead of the Ferranti Mark I, UNIVAC and LEO computers. Its first customer was the Technical University of Zurich.

Zuse's results are all the more impressive given that he was working alone in Germany, and he was unaware of the developments taking place in other countries. He had no university support or graduate students to assist him.

G. O'Regan, *Giants of Computing: A Compendium of Select, Pivotal Pioneers*, DOI 10.1007/978-1-4471-5340-5_60, © Springer-Verlag London 2013

Fig. 60.1 Konrad Zuse
(Courtesy of Horst Zuse,
Berlin)

60.1   Zuse's Machines

Zuse was unaware of computer-related developments in Germany or in other countries, and he independently implemented the principles of modern digital computers in isolation.

He commenced work on his first machine called the Z1 in 1936, and the machine was operational by 1938. It was demonstrated to a small number of people who saw it rattle and compute the determinant of a three by three matrix. It was essentially a binary electrically driven mechanical calculator with limited programmability. It was capable of executing instructions read from the program punched cards, but the program itself was never loaded into the memory.

It employed the binary system and metallic shafts that could slide from position 0 to position 1 and vice versa. The machine was essentially a 22-bit floating-point value adder and subtracter. A decimal keyboard was used for input, and the output was decimal digits. The machine included some control logic which allowed it to perform more complex operations such as multiplications and division. These operations were performed by repeated additions for multiplication and repeated subtractions for division. The multiplication took approximately 5 s. The computer memory contained 64 22-bit words. Each word of memory could be read from and written to by the program punched cards and the control unit. It had a clock speed of 1 Hz, and two floating-point registers of 22 bits each. The machine was unreliable and a reconstruction of it is in the Deutsches Technikmuseum in Berlin.

His next attempt was the creation of the Z2 machine which aimed to improve on the Z1. This was a mechanical and relay computer created in 1939. It used a similar mechanical memory but replaced the arithmetic and control logic with 600 electrical relay circuits. It used 16-bit fixed-point arithmetic instead of the 22-bit used in the Z1. It had a 16-bit word size and the size of its memory was 64 words. It had a clock speed of 3 Hz.

The Z3 machine was the first functional tape-stored-program-controlled computer and was created in 1941. It used 2,600 telephone relays and the binary number

Fig. 60.2 Zuse and the reconstructed Z3 (Courtesy of Horst Zuse, Berlin)

system and could perform floating-point arithmetic. It had a clock speed of 5 Hz and multiplication and division took 3 s. The input to the machine was with a decimal keyboard, and the output was on lamps that could display decimal numbers. The word length was 22-bits and the size of the memory was 64 words (Fig. 60.2).

It used a punched film for storing the sequence of program instructions. It could convert decimal to binary and back again. It was the first digital computer since it predates the Atanasoff-Berry Computer by 1 year. It was proven to be Turing complete in 1998. There is a reconstruction of the Z3 computer in the Deutsches Museum in Munich.

The Z4 was almost complete before the fall of Berlin to the advancing Red Army. Zuse escaped to Bavaria, and the completed Z4 machine was the world's first commercial computer when it was introduced in 1950.

60.2 Zuse and Plankalkül

The earliest high-level programming language was *Plankalkül* developed by Zuse in 1946. It means "Plan" and "Kalkül", i.e. a calculus of programs. It is a relatively modern language for a language developed in 1946. There was no compiler for the language, and the Free University of Berlin developed a compiler for it in 2000. This allowed the first Plankalkül program to be run over 55 years after its conception.

It employs data structures and Boolean algebra and includes a mechanism to define more powerful data structures. Zuse demonstrated that the Plankalkül language could be used to solve scientific and engineering problems, and he wrote several example programs including programs for sorting lists and searching a list for a particular entry. The main features of Plankalkül are:

- A high-level language.
- Fundamental data types are arrays and tuples of arrays.
- While construct for iteration.
- Conditionals are addressed using guarded commands.
- There is no GOTO statement.
- Programs are non-recursive functions.
- Type of a variable is specified when it is used.

The main constructs of the language are variable assignment, arithmetical and logical operations, guarded commands and while loops. There are also some list and set processing functions.

Chapter 61
Epilogue

This book has attempted to give a flavour of the work of a selection of those who have made important contributions to the computing field. It is not feasible, due to space constraints, to consider all those who merit inclusion. We gave a short account of each pivotal pioneer and included brief biographical information and a concise account of their contribution.

We discussed a selection of historical individuals who provided the foundation for the field. These included George Boole who developed Boolean algebra, which was later recognized by Claude Shannon as providing the appropriate mathematical model for the design of digital circuits. Charles Babbage designed the Difference Engine and the Analytic Engine. Lady Ada Lovelace worked with Babbage on applications of the Analytic Engine and is considered the world's first programmer. Gottfried Wilhelm Leibniz was a seventeenth-century German mathematician and inventor.

Herman Hollerith founded a tabulating company that would become International Business Machines (IBM). Vannevar Bush developed the first large-scale general-purpose mechanical analog computer at the Massachusetts Institute of Technology in the late 1920s. Claude Shannon showed how Boolean algebra could be applied to the design of digital circuits. The early computer pioneers built analog and digital computers.

Howard Aiken designed and built the Harvard Mark I computer which was an electromechanical calculator. John Atanasoff designed and built the Atanasoff-Berry Computer (ABC) digital computer in 1942. Tommy Flowers designed the Colossus computer at Bletchley Park in England in 1944. John Mauchly designed the ENIAC and EDVAC. John von Neumann gave his name to the fundamental architecture underlying computer systems. Alan Turing did important work in theoretical computing and on early computers at Bletchley Park. Sir Frederick Williams designed the Williams tube, which was used on the Manchester Mark I computer. Konrad Zuse is considered to be the "father of the computer" in Germany.

We discussed a selection of those who made important contributions to early commercial computing. These included Gene Amdahl, who did important work on early IBM computers, and he later formed Amdahl Corporation. John Backus

G. O'Regan, *Giants of Computing: A Compendium of Select, Pivotal Pioneers*,
DOI 10.1007/978-1-4471-5340-5_61, © Springer-Verlag London 2013

made important contributions to early programming languages. Gordon Bell was the architect for the PDP-4 and PDP-6 and VAX series of computers at Digital Corporation. Fred Brooks was the project manager for the IBM/360 project, and he later wrote the influential book *The Mythical Man Month*. Gordon Moore was one of the co-founders of the Intel Corporation, and he also formulated Moore's law. William Shockley and others developed the transistor at Bell Labs in the early 1950s.

We discussed a selection of individuals who made important contributions to later commercial computing. Vint Cerf and Bob Kahn invented the transmission control protocol (TCP) in the early 1970s. Edgar Codd developed relational databases at IBM. Don Estridge was the project manager for the team that developed the IBM personal computer. Gary Kildall wrote the first programs for the Intel 4004 microprocessor, and he developed the CP/M operating system for microprocessors. Tim Berners-Lee invented the World Wide Web, which has led to a revolution in the computing field.

We discussed several individuals who have made important contributions to software engineering. These include Robert Floyd who did important work on parsing and compilers in the 1960s and was one of the pioneers in investigating methods to prove the correctness of programs. C.A.R. Hoare developed the quicksort algorithm, axiomatic semantics of programming languages and the Calculus of Sequential Processes (CSP). Dines Bjørner and Cliff Jones developed the Vienna Development Method (VDM) at the IBM laboratory in Vienna. Edsger Dijkstra contributed to the development of graph algorithms, and his calculus of weakest preconditions is a methodology to develop a program and its proof of correctness hand in hand.

Tom DeMarco has made important contributions to project management and was one of the developers of structured analysis in the 1980s. Michael Fagan developed the Fagan Inspection methodology at IBM. Watts Humphrey made important contributions to software quality and to the development of process maturity models such as the Capability Maturity Model (CMM), PSP and TSP. Ivar Jacobson has made important contributions to UML and to the Rational Unified Process. David Parnas has made important contributions to the software field, and his ideas on the specification, design, implementation, maintenance and documentation of computer software remain important. Ed Yourdon has made contributions to systems analysis and design methodologies.

We discussed a selection of those who have made important contributions to theoretical computing and programming languages. These include Noam Chomsky who did important work on linguistics and grammars. Alonzo Church made important contributions to logic and computability. James Gosling developed the Java programming language at Sun Microsystems. Grace Murray developed the COBOL programming language. Kenneth Iverson developed the APL. Donald Knuth is considered to be the father of the analysis of algorithms, and he also developed the TeX and METAFONT typesetting systems. Dennis Ritchie developed the C programming language and codeveloped the UNIX operating system with Ken Thompson. Dana Scott made important contributions to the semantics of programming language and codeveloped the Scott-Strachey approach with Christopher

Strachey. Bjarne Stroustrup designed the C++ programming language, and Niklaus Wirth designed the Pascal programming language. Richard Stallman is the founder of the free software movement with the GNU project.

We discussed the field of artificial intelligence and considered individuals such as René Descartes, who formulated Cartesian dualism and made a clear distinction between mind and body. John McCarthy is considered the father of AI, and he believed that common-sense knowledge and reasoning can be formalized with logic. Marvin Minsky advocates a symbol manipulation approach as the centre of any attempt to understand intelligence. John Searle formulated the Chinese Room thought experiment which is a rebuttal of strong AI. Joseph Weizenbaum developed the ELIZA program, at MIT in 1966, and this famous program interacted with a user sitting at an electric typewriter in the manner of a psychotherapist.

We discussed a selection of computer entrepreneurs. This included figures such as Larry Ellison who founded Oracle Corporation, Bill Gates who founded Microsoft Corporation, Steve Jobs who founded Apple Corporation, Ken Olsen who founded Digital Corporation, and Thomas Watson Sr. and Jr. who were the former presidents of IBM.

61.1   Quo Vadimus?

The technological achievements in computing over the last 150 years are extraordinary. The human race has embarked on an amazing journey from the development of tabulators in the late nineteenth century to the development of analog computers, to the development of the early bulky digital computers, to the development of early commercial computers, to the development of transistors and integrated circuits, to the IBM mainframes and digital minicomputers, to the development of the microprocessor and home computers, to the rise of the IBM personal computer and the Apple Macintosh, to the rise of the Internet and World Wide Web, to the rise of the mobile phone, to the growth of Social Media, to the development of smartphones and so on.

We are living in a rapidly changing world where computer technology has driven innovation in almost all aspects of our world (e.g. communication, the media, medicine, automobiles, banking). These developments have led to major benefits to society, and it is natural to wonder where all of these innovations will lead. Will robots one day perform much of the work done by humans? Will real progress be made in the AI field? Will it be possible some day for machines to achieve human-like intelligence? Will technology assist in the elimination of poverty in the world? Will technology be used for good purposes? The pace of change is so relentless that any predictions made are likely to be wide of the mark. All that we can say is that it will be interesting.

Test Yourself (Quiz 1)

1. Explain the importance of Boole's equation $x^2 = x$ as a fundamental law of thought.
2. Describe how Boolean logic may be employed in the design of digital circuits.
3. Describe how Babbage's Difference Engine may be employed to compute polynomial functions.
4. Describe the components of the Analytic Engine and explain their function.
5. Explain why Lady Ada Lovelace is considered the world's first programmer and describe her ideas on applications of the Analytic Engine.
6. Explain why Shannon's Master's thesis is a key milestone in computing.
7. Explain why Shannon's work on Information Theory has been so influential.
8. Explain why Shannon is considered the father of modern cryptography.
9. Discuss the influence of Vannevar Bush on scientific research in the United States.
10. Describe how Hollerith's work on tabulators for the 1890 population census in the United States led to the birth of International Business Machines (IBM).
11. Describe Howard Aiken's work on the Harvard Mark I and its successors. Describe the applications of the Mark I.
12. Discuss the relevance of the Atanasoff-Berry Computer (ABC) and explain how it was ruled to be the first digital computer in the patent court case between Sperry and Honeywell.
13. Describe Flower's work on the Colossus computer at Bletchley Park and how the machine was used to crack the Lorenz codes.
14. Describe the components of the von Neumann architecture and explain each component.
15. Describe how Turing used the Turing machine to show that there is no mechanical procedure to decide the truth or falsity of any given mathematical problem.
16. Describe Turing's work at Bletchley Park during the Second World War.
17. Describe the Turing test and explain whether it is an appropriate test of machine intelligence.

G. O'Regan, *Giants of Computing: A Compendium of Select, Pivotal Pioneers*,
DOI 10.1007/978-1-4471-5340-5, © Springer-Verlag London 2013

18. Explain the importance of the Williams tube and describe how it was used on the Manchester "Baby" computer. Explain the importance of this machine in the history of computing.
19. Describe the machines developed by Zuse and explain the importance of the Z3 and Z4.
20. Describe the features of the Plankalkül language and explain why it took over 50 years for a Plankalkül program to be run.

Test Yourself (Quiz 2)

1. Discuss Amdahl's contributions to IBM and the Amdahl Corporation.
2. Explain Amdahl's law and its importance in parallel computing.
3. Describe Gordon Bell's contributions to DEC's PDP and VAX series of minicomputers.
4. Discuss the project management principles identified by Brooks from his experience as project manager of the IBM/360 project and subsequently recorded in the Mythical Man Month.
5. Explain why Brooks believes that there is no silver bullet that will guarantee project success with the project delivered on time, on budget and with the right quality.
6. Explain the significance of Moore's law.
7. Discuss the relevance of Shockley to the computing field.
8. Explain Iverson's views on the importance of notation as a tool of thought.
9. Describe Cerf's contributions to TCP/IP.
10. Describe Codd's relational model and SQL. Explain the importance of Codd's ideas.
11. Explain the relevance of Don Estridge in the history of computing. What errors did IBM make in the introduction of the IBM PC?
12. Explain why Gary Kildall has been described as the *man who could have been Bill Gates*.
13. Explain how Tim Berners-Lee invented the World Wide Web at CERN.
14. Explain the importance of Descartes to the AI field.
15. Discuss the feasibility of McCarthy's project, i.e. to develop programs with common-sense knowledge about the world.
16. Describe Searle's Chinese Room thought experiment and discuss whether it is an effective rebuttal of strong AI.
17. Describe Minsky's contributions to the AI field.

18. Explain how Weizenbaum became a leading critic of AI and explain his views on the ethics of AI.
19. Compare and contrast the achievements of Bill Gates and Steve Jobs.
20. What are your views on the management style of Ken Olsen and the Digital Corporation?
21. Compare and contrast the achievements of Thomas Watson Sr. and Jr.

Test Yourself (Quiz 3)

1. Explain the Chomsky hierarchy of grammars and how it is used in computer science.
2. Describe the Church-Turing thesis.
3. Describe Grace Murray Hopper's contribution to the COBOL programming language.
4. Describe Dana Scott's work in Denotation Semantics.
5. Explain why Donald Knuth is called the father of the analysis of algorithms.
6. Describe Wirth's work on Pascal and stepwise refinement.
7. Discuss the similarities and differences between C and C++.
8. Discuss the importance of the Java programming language.
9. Compare and contrast the approaches of Hoare, Dijkstra and Parnas.
10. Describe Fagan Inspections. How effective are they in building quality into the software?
11. Explain why Watts Humphrey is known as the father of software quality.
12. Discuss the contributions of DeMarco and Yourdon.

G. O'Regan, *Giants of Computing: A Compendium of Select, Pivotal Pioneers*,
DOI 10.1007/978-1-4471-5340-5, © Springer-Verlag London 2013

Glossary

ABC Atanasoff-Berry Computer
ACE Automatic Computing Engine
ACM Association for Computing Machinery
ADEC Aiken Dahlgren Electronic Calculator
AI Artificial intelligence
ALGOL Algorithmic language
AOL America Online
APL A Programming Language
ASCC Automatic Sequence Controlled Calculator
ASCII American Standard Code for Information Interchange
AT&T American Telephone and Telegraph Company
B2B Business-to-Business
B2C Business-to-Consumer
BASIC Beginners All-Purpose Symbolic Instruction Code
BIOS Basic Input/Output System
BNF Backus-Naur Form
CCITT Comité Consultatif International Téléphonique et Télégraphique
CEO Chief executive officer
CERN Conseil European Recherche Nucleaire
CMM® Capability Maturity Model
CMMI® Capability Maturity Model Integration
COBOL Common Business Oriented Language
CODASYL Conference on Data Systems Languages
CP/M Control Program for Microcomputers
CPSR Computer Professionals Social Responsibility
CPU Central processing unit
CRM Customer relationship management
CSIRAC Council for Scientific and Industrial Research Automatic Computer
CSP Communication Sequential Processes
CTR Computing Tabulating Recording Company
DARPA Defence Advanced Research Project Agency

G. O'Regan, *Giants of Computing: A Compendium of Select, Pivotal Pioneers*,
DOI 10.1007/978-1-4471-5340-5, © Springer-Verlag London 2013

DB Database
DBA Database administrator
DBMS Database management system
DDL Data Definition Language
DEC Digital Equipment Corporation
DFD Data flow diagram
DML Data Manipulation Language
DNS Domain naming system
DOS Disc operating system
DRAM Dynamic random access memory
DRI Digital Research Incorporated
DVD Digital Versatile Disc
EDVAC Electronic Discrete Variable Automatic Computer
EMCC Eckert-Mauchly Computer Corporation
ENIAC Electronic Numerical Integrator and Computer
ERD Entity relationship diagrams
ERP Enterprise resource planning
ETH Swiss Federal Institute of Technology, Zurich
FTP File transfer protocol
FSF Free Software Foundation
FSM Finite state machine
GECOS General Electric Comprehensive Operating System
GM General Motors
GNU GNU's Not Unix
GUI Graphical user interface
HP Hewlett-Packard
HTML Hypertext markup language
HTTP Hypertext transport protocol
IBM International Business Machines
IEEE Institute of Electrical and Electronic Engineers
IP Internet protocol
ISO International Standards Organization
ITU International Telecommunications Union
JCP Java Community Process
JDK Java Development Kit
JIT Just in time
JSP Jackson Structured Programming
JVM Java Virtual Machine
LEO Lyons Electronic Office
LSD Lysergic acid diethylamide
LSI Large-scale integration
LT Logic theorist
MADC Manchester Automatic Digital Computer
MIMD Multiple instruction, multiple data
MIPS Million instructions per second

MIT Massachusetts Institute of Technology
MITS Micro Instrumentation and Telemetry System
MS-DOS Microsoft Disc Operating System
MSN Microsoft Network
NACA National Advisory Committee Aeronautics
NASA National Aeronautics and Space Administration
NCR National Cash Register
NDRC National Defence Research Committee
NFA Non-deterministic Finite State Automata
NPL National Physical Laboratory
NR Norwegian Research
OMG Object Management Group
OMT Object Modeling Technique
OOSE Object-oriented software engineering
OS Operating system
OSRD Office Scientific Research and Development
PC Personal computer
PDA Personal digital assistant
PDP Programmed Data Processor
PL/M Programming Language for Microcomputers
PS/2 Personal System 2
RAISE Rigorous Approach to Industrial Software Engineering
RAM Random access memory
RDBMS Relational database management system
RSL RAISE Specification Language
RUP Rational Unified Process
SAGE Semi-Automatic Ground Environment
SDI Strategic Defence Initiative
SDL Specification and Description Language
SEI Software Engineering Institute
SMTP Simple Mail Transfer Protocol
SNARC Stochastic Neural Analog Reinforcement Computer
SQL Structured Query Language
SQL/DS Structured Query Language/Data System
SSEC Selective Sequence Electronic Computer
SSEM Small-scale experimental machine
TCP Transport control protocol
UCLA University of California (Los Angeles)
UDP User datagram protocol
UG Universal grammar
UML Unified Modelling Language
UNIVAC Universal Automatic Computer
URL Universal resource locator
VAX Virtual Address eXtension
VDM Vienna Development Method

VDM♣ Irish School of VDM
VLSI Very-large-scale integration
VM Virtual memory
VMS Virtual memory system
W3C World Wide Web Consortium
WISC Wisconsin Integrally Synchronized Computer

References

[AGH:13]   Arnold K, Gosling J, Holmes D (2013) The java programming language, 5th edn. Prentice Hall, New York
[Amd:67]   Amdahl G (1967) Validity of the single processor approach to achieving large-scale computing capabilities. In: AFIPS conference proceedings, vol 30. Thompson Publishing, Washington, DC, pp 483–485
[Bab:42]   Menabrea LF (1842) Sketch of the analytical engine. Invented by Charles Babbage (trans: Lovelace LA). Bibliothèque Universelle de Genève
[Bac:53]   Backus J (1953) The IBM 701 speedcoding system. International Business Machines, New York
[Bac:78]   Backus J (1978) Can programming be liberated from the von Neumann style? Turing award lecture (1977). Commun ACM 21(8):613–641
[BjJ:78]   Bjørner D, Cliff J (1978) The Vienna development method. The meta language, vol 61, Lecture notes in computer science. Springer, New York
[BjJ:82]   Bjørner D, Cliff J (1982) Formal specification and software development, Prentice Hall international series in computer science. Prentice Hall, Englewood Cliffs
[BL:00]   Berners-Lee T (2000) Weaving the web. Collins Book, San Francisco
[Blo:04]   Bloomberg (2004) The man who could have been Bill Gates. Bloomberg Business Week Magazine, October 24th 2004
[Boe:81]   Boehm B (1981) Software engineering economics. Prentice Hall, Englewood Cliffs
[Boo:48]   Boole G (1848) The calculus of logic. Camb Dublin Math J III:183–198
[Boo:58]   Boole G (1958) An investigation into the laws of thought. Dover Publications, New York. (First published in 1854)
[Brk:75]   Brooks F (1975) The mythical man month. Addison Wesley, Reading
[Brk:86]   Brooks F (1986) No silver bullet. Essence and accidents of software engineering, Information processing. Elsevier, Amsterdam
[Bus:45]   Bush V (1945) As we may think. Atlantic Mon 176(1):101–108
[CKS:11]   Chrissis MB, Conrad M, Shrum S (2011) CMMI. Guidelines for process integration and product improvement, 3rd edn, SEI series in software engineering. Addison Wesley, Boston
[Cod:70]   Codd EF (1970) A relational model of data for large shared data banks. Commun ACM 13(6):377–387
[Dat:81]   Date CJ (1981) An introduction to database systems, 3rd edn, The systems programming series. Addison-Wesley, Reading
[DeM:09]   DeMarco T (2009) Software engineering. An idea whose time has come and gone? IEEE Softw 26(4):94–95
[DeM:79]   DeMarco T, Plauger PJ (1979) Structured analysis and system specification. Prentice Hall, Englewood Cliffs

G. O'Regan, *Giants of Computing: A Compendium of Select, Pivotal Pioneers*,
DOI 10.1007/978-1-4471-5340-5, © Springer-Verlag London 2013

[DeM:86]   DeMarco T (1986) Controlling software projects: management, measurement, estima-
           tion. Prentice Hall, Upper Saddle River
[DeM:99]   DeMarco T, Lister T (1999) Peopleware: productive projects and teams, 2nd edn.
           Dorset House Publishing, New York
[Des:99]   Descartes R (1999) Discourse on method and meditations on first philosophy, 4th edn
           (trans: Cress D). Hackett Publishing Company, New Haven
[Dij:68]   Dijkstra EW (1968) Go to statement considered harmful. Commun ACM 11(3):
           147–149
[Dij:72]   Dijkstra EW, Hoare CAR, Dahl OJ (1972) Structured programming. Academic, New
           York
[Dij:76]   Dijkstra EW (1976) A disciple of programming. Prentice Hall, Englewood Cliffs
[Fag:76]   Fagan M (1976) Design and code inspections to reduce errors in software develop-
           ment. IBM Syst J 15(3):182–210
[Fen:95]   Fenton N (1995) Software metrics: a rigorous approach. Thompson Computer Press,
           London
[Flo:63]   Floyd R (1963) Syntactic analysis and operator precedence. J Assoc Comput Mach
           10:316–333
[Flo:64]   Floyd R (1964) The syntax of programming languages. A survey. IEEE Trans Electron
           Comput EC-13:346–353
[Flo:67]   Floyd R (1967) Assigning meanings to programs. In: Proceedings of symposia in
           applied mathematics, vol 19. American Mathematical Society, Providence, pp 19–32
[Glb:94]   Gilb T, Graham D (1994) Software inspections. Addison Wesley, New York
[Glb:76]   Gilb T (1976) Software metrics. Winthrop Publishers, Inc., Cambridge
[Gri:81]   Gries D (1981) The science of programming. Springer, Berlin
[Hen:80]   Heninger K (1980) Specifying software requirements for complex systems. IEEE
           Trans Softw Eng 6(1):2–13
[Hor:69]   Hoare CAR (1969) An axiomatic basis for computer programming. Commun ACM
           12(10):576–585
[Hor:85]   Hoare CAR (1985) Communicating sequential processes, Prentice Hall international
           series in computer science. Prentice Hall, Englewood Cliffs
[Hum:89]   Humphrey W (1989) Managing the software process. Addison Wesley, Reading
[InA:91]   Darrell I, Derek A (1991) Practical formal methods with VDM, McGraw Hill
           international series in software engineering. McGraw Hill, New York/London
[Ive:62]   Iverson K (1962) A programming language. Wiley, New York
[Jac:92]   Jacobson I (1992) Object-oriented software engineering: a use case driven approach.
           Addison Wesley, Reading
[Jac:05]   Jacaobson I et al (2005) The unified modelling language, user guide, 2nd edn. Addison
           Wesley Professional, Reading
[Jac:99]   Jacaobson I et al (1999) The unified software development process. Addison Wesley,
           Reading
[Jan:97]   Janicki R (1997) On a formal semantics of tabular expressions. Technical report CRL
           355. Communications Research Laboratory. McMaster University, Hamilton
[KaC:74]   Cerf V, Kahn B (1974) Protocol for packet network interconnections. IEEE Trans
           Commun Technol 22(5):637–648
[KeR:78]   Kernighan B, Ritchie D (1978) The C programming language, 1st edn, Prentice Hall
           software series. Prentice Hall, Reading
[Knu:79]   Knuth DE (1979) Mathematical typography. Bull Am Math Soc 1(2):337–372
[Lei:03]   Leibniz GW (1703) Explication de l'Arithmétique Binaire. Memoires de l'Academie
           Royale des Sciences
[Lov:42]   Menabrea LF (1842) Sketch of the analytic engine invented by Charles Babbage.
           Bibliothèque Universelle de Genève, no. 82 (trans: Ada, Augusta, Countess of
           Lovelace)

[Mc:59]   McCarthy J (1959) Programs with common sense. In: Proceedings of the Teddington conference on the mechanization of thought processes. Her Majesty's Stationery Office, London

[Mc:60]   McCarthy J (1960) Recursive functions of symbolic expressions and their computation by machine. Commun ACM 3(4):184–195

[McH:85]   McHale D (1985) Boole. Cork University Press, Cork

[Min:61]   Minsky M (1961) Steps towards artificial intelligence. In: Feigenbaum E, Feldman J (eds) Computers and thought. McGraw Hill, New York, pp 406–450

[Min:65]   Minsky M (1965) Matter, minds and models. In: Proceedings of the international federation of information processing congress, vol 1, pp 45–49

[Min:74]   Minsky M (1974) A framework for representing knowledge. MIT-AI Laboratory Memo 306, Cambridge

[Min:79]   Minsky M (1979) A theory of memory. MIT-AI Laboratory Memo 516, Cambridge

[Min:88]   Minsky M (1988) The society of mind. Simon and Schuster Press, New York

[MiP:69]   Minsky M, Papert S (1969) Perceptrons. MIT Press, Cambridge, MA

[Mor:65]   Moore G (1965) Cramming more components onto integrated circuits. Electronics Magazine, pp 114–117, April 19

[Mun:10]   Munson GE (2010) The rise and fall of Unimation Inc. A story of robotics innovation and triumph that changed the world. Robot 24 (September– October)

[Nyq:24]   Nyquist H (1924) Certain factors affecting telegraph speed. Bell Syst Tech J 3: 324–346

[ORg:02]   O'Regan G (2002) A practical approach to software quality. Springer, New York

[ORg:06]   O'Regan G (2006) Mathematical approaches to software quality. Springer, London

[ORg:10]   O'Regan G (2010) Introduction to software process improvement. Springer, London

[ORg:12]   O'Regan G (2012) A brief history of computing, 2nd Edition. Springer, London

[ORg:13]   O'Regan G (2013) Mathematics in computing. Springer, London

[Par:92]   Parnas DL (1992) Tabular representation of relations. CRL report 260. McMaster University, Canada

[Rus:45]   Russell B (1945) A history of western philosophy. George Allen and Unwin Ltd., London

[Sch:86]   Schmidt D (1986) Denotational semantics. A methodology for language development. Allyn and Bacon, Boston

[Sea:80]   Searle J (1980) Minds, brains, and programs. Behav Brain Sci 3:417–457

[SEI:06]   Software Engineering Institute (2006) CMMI executive overview. Presentation by the SEI

[Sha:37]   Shannon C (1937) A symbolic analysis of relay and switching circuits. Masters thesis. Massachusetts Institute of Technology

[Sha:48]   Shannon C (1948) A mathematical theory of communication. Bell Syst Techn J 27:379–423

[Sha:49]   Shannon CE (1949) Communication theory of secrecy systems. Bell Syst Tech J 28(4):656–715

[Sho:50]   Shockley W (1950) Electrons and holes in semiconductors with applications to transistor electronics. Van Nostrand, New York

[Sta:02]   Stallman RM (2002) Free software, free society, 2nd edn. Free Software Foundation, Inc., Boston

[Sto:77]   Stoy JE (1977) Denotational semantics: the Scott-Strachey approach to programming language semantics. MIT Press, Cambridge, MA

[Tur:50]   Turing A (1950) Computing, machinery and intelligence. Mind 49:433–460

[VN:32]   von Neumann J (1932) Mathematische Grundlagen der Quantenmechan (The mathematical foundations of quantum mechanics). Springer, Berlin

[VN:45]   von Neumann J (1945) First draft of a report on the EDVAC. University of Pennsylvania, Philadelphia

[Wei:66]   Weizenbaum J (1966) ELIZA. A computer program for the study of natural language communication between man and machine. Commun ACM 9(1):36–45

[Wei:76]   Weizenbaum J (1976) Computer power and human reason: from judgments to calculation. W.H. Freeman and Co Ltd, San Francisco

[Whi:11]   Whitehead AN (1911) Introduction to mathematics. H. Holt and Company, New York

[Wir:71]   Wirth N (1971) Program development by stepwise refinement. Commun ACM 14(4):221–227

Index

306   Index

Printed in the United States
By Bookmasters